Contents

introduction to

ENERGY, ENVIRONMENT, SUSTAINABILITY

Second Edition

paul gannon
montana state university

Kendall Hunt
publishing company

Kendall Hunt
publishing company

www.kendallhunt.com
Send all inquiries to:
4050 Westmark Drive
Dubuque, IA 52004-1840

Printed in the United States of America
10 9 8 7 6 5 4 3 2 1

Preface

Human society has never before faced challenges like the present. Population has more than doubled since 1965; currently about 7 billion and growing fast. Much of this growth occurs in regions with developing infrastructure, placing increasing demands on natural resources and taxing ecosystem services. Coupled with continued growth in more developed, and more demanding regions, global energy consumption is projected to increase nearly *three-fold* by mid-century. At present, over 80% of global energy demand is supplied by fossil fuels (coal, oil, and natural gas), whose abundance and energy density has enabled population growth beyond ecosystem capacity. The burning of fossil fuels is primarily responsible for the increasingly rapid global warming observed in the past century. This anthropogenic (human-caused) warming has destabilized regional climates, affecting living systems around the planet, threatening food and water supplies and increasing the frequency and intensity of severe weather events. Despite this knowledge, fossil fuels are in record high demand, and their supply, distribution, and use powers a globalizing economy. Moreover, the locations and amounts of the limited coal, oil, and gas reserves around the planet are geographically uneven, which fuels many regional and global conflicts. These interconnected challenges are often referred to as the "energy, environment, prosperity dilemma".

Although very serious, these challenges are not unsolvable. Opportunities to adapt and progress toward an enduring globally-equitable society are at our fingertips. As society comes to appreciate the urgency in confronting this dilemma, strategies will be implemented to adapt to climate change pressures, and secure water, food, and energy systems. Concurrently, development of low-carbon and renewable energy sources will accelerate. Additionally, carbon sequestration (from the atmosphere) will be increased through reforestation, thoughtful agricultural practices and perhaps new technologies. Human society will thus continue endeavors to realize the elusive ideal of sustainability. Essential in these efforts is basic education to promote public literacy of energy, water, and food infrastructures, as well as environmental sciences and sustainability.

This textbook was originally created to facilitate an entry-level university science course (ECHM 205CS: Energy and Sustainability), first offered by the Chemical Engineering Department in 2009 at Montana State University (MSU) in Bozeman, MT. The course surveys a range of contemporary scientific and technological topics in a manner intended to be accessible to non-science/engineering students, while encouraging critical thinking and communication skills. At MSU, the course also provides University-Core credit for a "Contemporary Issues in

Science" graduation requirement, and is often the only science course taken by some undergraduate students. The course and textbook aim to be clear, concise and interactive, providing a foundation for understanding contemporary issues and encouraging students to explore and engage in science and in their communities. At MSU, student groups are formed to study a specific course-related topic and then present posters at a public event near the end of the term. Many of these groups interact with individuals and organizations within the community to strengthen their presentations.

The text is organized into ten (10) sequential chapters and is designed for a single academic term. Chapters 1–3 present an overview of human society and its impacts, as well as energy and environmental sciences and Earth System dynamics. Chapter 4 reviews the basics of combustion (fire) as well as its utility and globalized impacts since the Industrial Revolution, focusing on global climate destabilization. Chapter 5 discusses non-renewable energy sources (fossil fuels) and related exploration, production, and conversion technologies. Chapter 6 covers atomic energy basics and nuclear energy technologies. Chapter 7 overviews renewable energy sources and conversion technologies. Chapter 8 introduces basic concepts of electricity and hydrogen. Chapter 9 considers contemporary issues surrounding food and water systems. Chapter 10 concludes with reflections on science, sustainability, and globalizing human society. End-of-chapter quizzes and problems aim to facilitate comprehension, and can be easily adapted for homework assignments and exams.

Accompanying this second edition for instructors is an updated website with sample course syllabi, lecture slides, solution to end-of-chapter quizzes/problems, suggested classroom demonstrations and activities, and sample multiple-choice exams.

We will continue improving this textbook with user feedback—thanks for your help!

–Paul Gannon
pgannon@montana.edu

About the Author

Paul grew up in Montana. He completed his undergraduate and graduate studies in Chemical Engineering at Montana State University (MSU), and has been an Assistant Professor in Chemical Engineering at MSU since 2008. His research and teaching focuses on sustainable energy technologies, and he is active in campus sustainability initiatives. Paul also founded and directs a research laboratory where students study the behaviors of high-temperature materials used in silicon manufacturing, fuel cell systems, and turbine engines, with work funded by private industry as well as by various State and Federal agencies, including NASA and the US Department of Energy. The laboratory has produced over two dozen peer-reviewed journal articles describing the research, and Paul enjoys international collaboration with colleagues in industry, national laboratories, and academia.

Paul is also passionate about teaching and learning, and has instructed both new and old courses at MSU, including Materials Science, Thermodynamics, and Energy and Sustainability, which motivated this textbook. He has taught over 1200 MSU undergraduate students in the classroom, and has worked with several dozen in the laboratory. Paul has also had the pleasure of working with and learning from numerous graduate and post-graduate students. Student success in the classroom, laboratory, and beyond continues to drive his work and provides him hope for the future.

Paul also enjoys many forms of outdoor recreation and socializing with friends and family. Along with his beautiful spouse (and accomplished artist) Cherlyn Wilcox and their faithful dog, Zeppelin, Paul lives, works, and plays in and around Bozeman, MT, about 50 miles north of Yellowstone Park.

chapter 1

Human Society and Ecological Footprint

"If the world is to save any part of its resources for the future, it must reduce not only consumption but the number of consumers."

— Psychologist and author B.F. Skinner, Introduction to Walden Two, 1976

Origins, Evolution and Revolutions

Anatomically-modern humans (homo sapiens) evolved from our pre-human ancestors around 130,000–170,000 years ago. For tens of thousands of years, humans developed primitive hunter-gatherer (or forager) societies throughout the forests and grasslands of what is now eastern Africa. Around 60,000–70,000 years ago, some humans began migrating out of Africa into what is now the Middle East, Europe, Asia, and ultimately the Americas. By analyzing genetic data from people around the world, specific mutations can be used to trace genetic lineages, all the way back to the first humans in Africa. **Figure 1.1** approximates global human migrations throughout time (years before present) from our origins in Africa, based on these genetic data.[1]

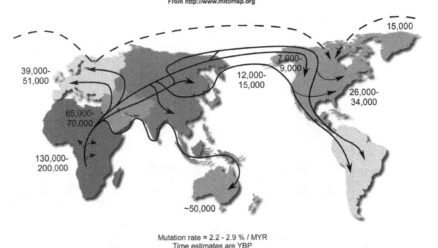

Figure 1.1 Estimated human migrations on Earth as determined by analysis of genetic mutations (mitochondrial DNA).[1]

Human populations were historically small and isolated. The hunter-gatherer societies could support significant populations (>100 individuals), yet were vulnerable to ecosystem changes. In rich ecosystems during good climates, hunter-gatherer societies enjoyed near modern life-spans and reasonable labor requirements, permitting plenty of time for social and cultural development. During bad climates and ecosystem collapse, hunter-gather societies were forced to move or perish, presumably creating resource conflicts among more-densely populated regions. About 10,000 years ago, as the last ice-age subsided, the climate became relatively warm and mild—conditions that have persisted to the present. During this time, agriculture began appearing in human populations around the world, and many transitioned into *agrarian* (agriculture-based) *societies*. This marks the beginning of the Agricultural Revolution in human history.

Agrarian societies grew substantial populations (>1,000 individuals). This was partly due to agriculture producing more reliable food than hunting-gathering, but also because agriculture required significantly more labor. Additionally, the agrarian lifestyle typically afforded higher life-expectancy at birth and increased longevity. Eventually, the storage of grains helped insulate societies from ecosystem changes, such as drought, and help humans discover new agricultural products, such as beer. Populations became more centralized, and labor specialization and trade began to change human society forever. Numerous language-based cultures emerged from these social systems and grew populations with established territories and governance. However, these societies were also vulnerable to diseases, crop failures and war, all of which occasionally devastated populations. From about 10,000BCE (Before the Common Era—same as "BC"), agrarian societies slowly began to dominate the global human population, and humans evolved socially and culturally through various "ages" associated with the contemporary materials used for tools, such as stones, bronze, and iron. Hunter-gatherer societies also continued to evolve socially and culturally, and exist to this day in isolated pockets around the world. Some groups even today remain uncontacted (and undisturbed) by the modern world.

Early on, total human population on Earth was as low as a few thousand individuals, which grew into several million before the Agricultural Revolution. Following the Agricultural Revolution, global human population grew steadily to over a hundred million individuals by the dawn of the Common Era (CE—same as "AD"). As major human civilizations around the world grew, the population continued to increase, reaching the 1 billion (1,000 million) individual milestone in about the year 1800CE. Coincidentally, this was also the beginning of the Industrial Revolution, which permitted a population explosion by leveraging the extremely-dense energy source of coal. Coal was burned to boil water into steam, and the steam was used to drive engines to power modern civilization and help grow human population. The first practical coal-fired steam engines were developed in the late 1700s in the UK, and originally used for pumping water, often from coal mines. The utility of coal-powered devices quickly sp read, and the Industrial Revolution was underway.

While it required over 10,000 years for the global population to reach 1 billion, the next billion was added in a little over a century, doubling the population to 2 billion by about 1930CE. The next billion only required about two more decades, reaching 3 billion by about 1960CE. Then, within only four more decades, the population again doubled, reaching 6 billion in about 2000CE. Just over a decade later, the global population is over 7 billion individuals, living in over 190 countries and speaking over 7,000 languages throughout the world. **Figure 1.2** illustrates the history of global human population on Earth, with major ages indicated, and population predicted to reach over 8 billion individuals by 2025CE.

Human population projections from the United Nations are presented in **Figure 1.3**. The medium projection anticipates about 9 billion people by about 2050, with a leveling and decrease in population before the end of the century. These projections are based upon various assumptions of growth within both developed and developing countries. The vast majority of population increases are anticipated in the developing countries, including China, India, and various African and South American nations.

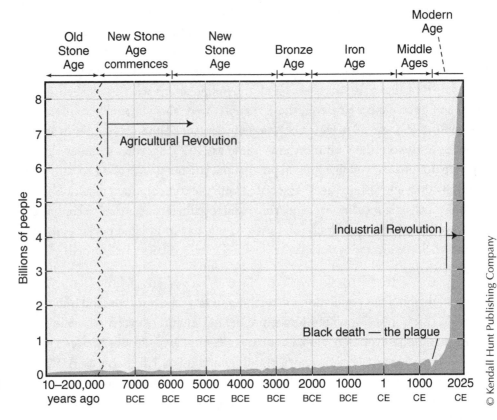

Figure 1.2 Human population over 200,000 years.

© Kendall Hunt Publishing Company

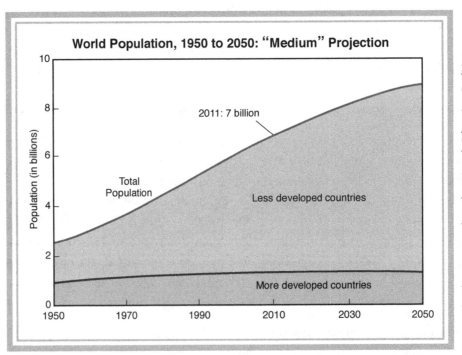

Figure 1.3 Human population projections to 2050.

Reprinted with permission from the Minerals Education Coalition. Adapted from John Christensen-Teri Christensen, *Global Science: Earth/ Environmental Systems Science, 7th Edition* (Dubuque, IA: Kendall Hunt Publishing Company, © 2009).

By continent, Asia is the most populated, and growing at a rate of about 12% over the past decade. Africa is a distant second, but has the highest growth (26% over the decade). The European continent has the third highest population, which has been about steady (not growing or declining) within the past decade. By country, China and India are the clear leaders in population, with over a billion people in each. The United States is a distant third with about 300 million people. Overall, the largest populations and growth rates are in the developing regions of the world. Access to education, energy, and health care is often correlated with lower population growth rates.

Modern Day Amenities

Humans in the developed regions of the world enjoy a myriad of modern amenities beyond the necessities of food, water, and shelter. These range from transportation and lighting, to comfortable accommodations, hot showers, cold beverages, and numerous goods and services that improve our lives. Here we discuss a few of these amenities and how they support modern civilization.

Transportation

Humans have used wind, water, and animals to aide in transportation for thousands of years. Following the Industrial Revolution and the roll out of trains, planes, ships, and automobiles in the early 1900's, transportation took on a new face. Transportation today is predominately powered by burning liquid fossil fuels (derivatives from crude oil), which drives the combustion engines powering most transportation systems, as illustrated in **Figure 1.4**. A smaller, but growing sector of transportation is powered by electricity, which can be derived from both renewable and non-renewable energy sources, as discussed later. An even smaller, but also growing sector of transportation relies on human power, such as walking or cycling.

Figure 1.4 Modern forms of transportation.

Image © Kar, 2012. Used under license from Shutterstock, Inc.

Lighting

In the developed regions of the world we often take lights and lighting for granted. Lighting, or illumination, is used when we do not have natural light, and/or for entertainment. Light is comprised of radiant (electromagnetic) energy, typically from the conversion of electrical energy (electricity). The characteristics of the light (radiant energy) depend on the materials involved in the conversion process. For example, when electricity flows through a tungsten wire in a glass bulb, it reaches very high temperatures and glows—this is called incandescence. Incandescent light bulbs (lamps) are only ~5% efficient in converting electricity into light; the remaining electricity is converted into heat (light bulbs are often hot to the touch).

When electricity flows through a bulb containing a gas, like mercury, it can produce different colors of light depending on the gas composition and/or phosphor coatings on the bulbs. These fluorescent light bulbs illuminate most large buildings and many homes with the increasingly popular compact fluorescent lights (CFLs). Fluorescent lighting is ~25% efficient in converting electricity into light. Light emitting diodes (LEDs) make use of semiconductor materials converting electricity directly into light (electroluminescence). Although currently more expensive, LEDs use about 10 times less energy than incandescent bulbs and have lifetimes exceeding both incandescent and CFLs. Various common small-scale light bulbs; incandescent, CFL and LED are presented in **Figure 1.5**.

Figure 1.5 Common light bulbs; incandescent, CFL and LED.

Image © maxstockphoto, 2012. Used under license from Shutterstock, Inc.

Heating and Cooling

Another modern amenity we often take for granted is heating and cooling within the buildings or vehicles we occupy. In a related manor, refrigeration and freezing has enabled critical and extended storage for stocks of food and medicine.

In colder regions, we are commonly concerned with hot water and hot air, which in our homes is typically provided by hot water heaters and air furnaces appliances. Often, we use electricity and/or gas to power these appliances, as shown in **Figure 1.6**. The gas is burned using air from the room and heats up the cold water or cold air that is passed over the burner, with the exhaust vented from a roof-top pipe. From the water heater, hot water is then distributed throughout the home by pipes, which deliver it to faucets or showerheads for hot water on-demand. Similarly, the air furnace delivers hot air throughout the house by air ducts, which deliver it to vents within rooms. Electricity can be used in lieu of gas by simply passing cold water or air over electrically-heated filaments.

Both heating and cooling operations may take advantage of *phase changes* in materials, like ice melting into water and water boiling into steam. Whenever *phase changes* take place, thermal energy is either taken from or released to the local environment. **Figure 1.7** pictures *phase changes* of water and associated energy absorbing or releasing effects. When water vapor condenses into liquid, it releases the energy required to change its phase from liquid to vapor. This is called *latent heat* ("latent" means "hidden" in Latin) of vaporization, and can be realized by carefully placing your hand above a pot of boiling water; when the steam condenses, it rapidly heats your hand. Conversely, when liquid evaporates it absorbs energy from the environment. This is realized when blowing on wet vs. dry skin; when the water evaporates into vapor it takes energy from your hand making it feel cool.

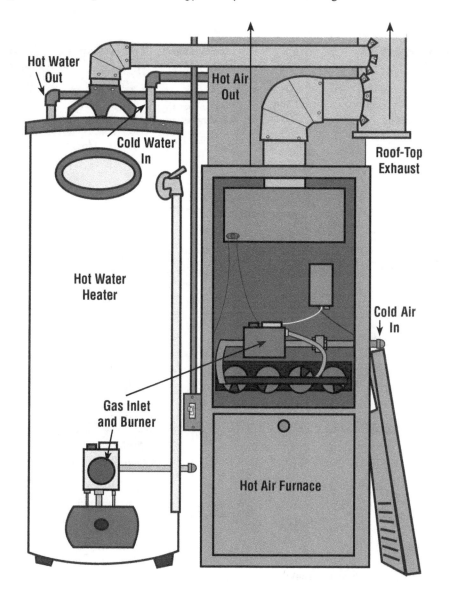

Image © Dennis Cox, 2012. Used under license from Shutterstock, Inc.

Figure 1.6 Common features of gas-powered water heater and air furnace as home appliances.

Like water, other fluids undergo phase changes at various pressures and temperatures and can be used as refrigerant fluids. The process of evaporation and condensation of a refrigerant fluid can be contained in a closed-loop system, with the refrigerant fluid and operation pressure chosen for the desired temperatures for evaporation and condensation. Presented in **Figure 1.8**, the refrigerant fluid undergoes a 4-step cycle of:

1. being compressed from a vapor into mostly liquid by a mechanical compressor (typically powered by electricity);

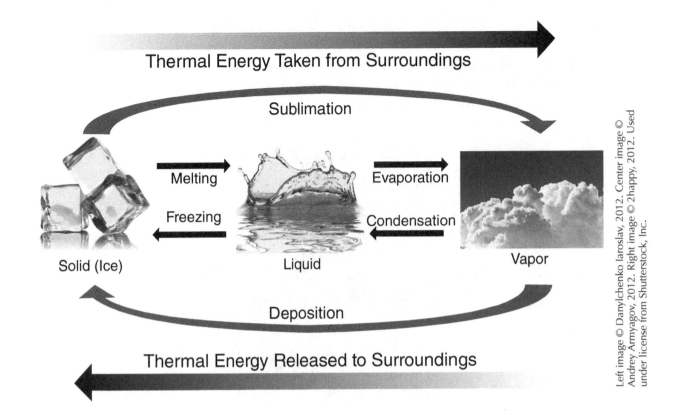

Left image © Danylchenko Iaroslav, 2012. Center image © Andrey Armyagov, 2012. Right image © 2happy, 2012. Used under license from Shutterstock, Inc.

Figure 1.7   Phase changes and associated energy release or absorption from environment.

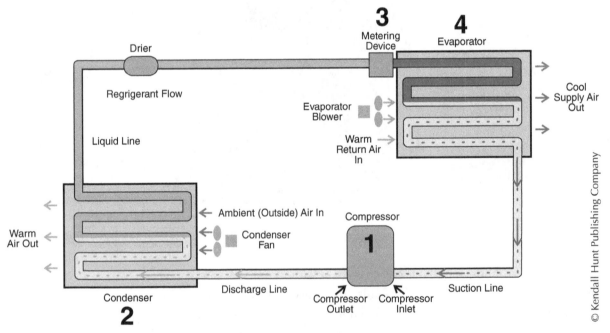

© Kendall Hunt Publishing Company

Figure 1.8   Basic features of a vapor-compression refrigeration cycle.

2. rejecting energy and condensing fully into a high pressure liquid in the condenser;

3. being metered into an evaporator by a valve; and,

4. evaporating from a liquid into a vapor and absorbing energy from the environment, making it cool.

This vapor-compression refrigeration cycle essentially pumps thermal energy from one region to another to effect both heating and cooling. In a refrigerator, the compressor and condenser reside outside of the cold zone, and while it does feel cold when you open the refrigerator door, the thermal energy is only being transferred to the back of the refrigerator. This is why refrigerators cannot effectively cool a room (unless the compressor and condenser are outside). Refrigeration cycles also often are integrated into larger heating, ventilation, and air-conditioning (HVAC) systems, which blow hot or cold filtered air within houses and buildings. Sophisticated HVAC systems integrated with lighting systems and active controls can save a tremendous amount of energy in buildings, which presently consume about a third of all energy in the United States. A simplified schematic of an HVAC system is shown in **Figure 1.9**.

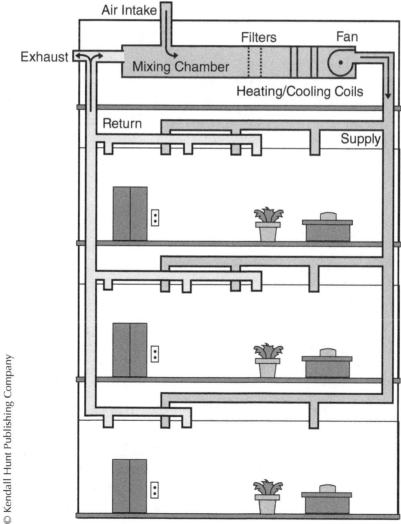

Figure 1.9 Basic features of a building HVAC system.

Stuff

In the developed regions and increasing in developing regions, we humans consume a lot of *stuff*. Here, *stuff* is intended to mean all of the numerous goods and services that we use regularly. *Stuff* includes clothes, clean water and toilets, and electricity along with all of the *stuff* it powers, like washing and drying machines, computers, as well as the i-phones, i-pods and i-pads that many of us enjoy. All this *stuff* requires energy and materials to make, transport, sell, use, and dispose of *stuff*, so consuming *stuff* causes environmental impacts. When most of the over 7 billion individuals consume *stuff*, the impacts are compounded. Next we discuss how to track the impacts of *stuff*.

Life-cycle Analysis

Life-cycle analysis (LCA) or life-cycle assessment is a way of looking at any product to determine its energy requirements and environmental impacts during its production, use, and disposal—its whole life-cycle, as illustrated in **Figure 1.10**. For this reason, LCA is often referred to as *cradle-to-grave* or *cradle-to-cradle* analysis if recycling is involved. LCA generally involves four steps:

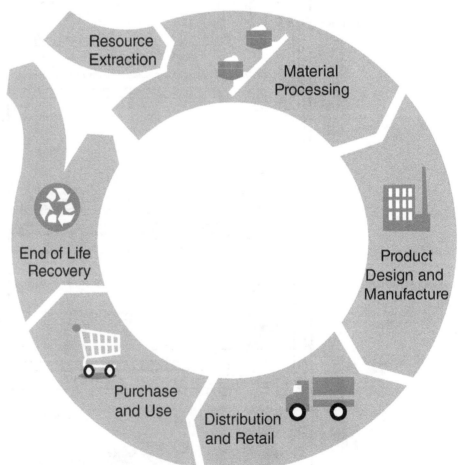

Figure 1.10 Basic components of a life-cycle assessment (LCA).

Source: Environmental Protection Agency.

1. determine a clear goal and scope for the LCA, i.e., determine the system boundaries and identify critical information;

2. compile an inventory of energy and material inputs and outputs at each stage of the life-cycle (commonly materials, manufacturing, packaging, distribution, usage and disposal—Figure 1.10);

3. evaluate the impacts of the specific energy and materials inputs and output; and,

4. interpret and communicate the results to inform product decisions.

LCA is an important tool to help quantify environmental, social, and economic impacts of products and processes, and is being used by an increasing number of organizations around the world. For example, if we were to perform a simple LCA for a text book, we might consider how/where the book was made, as well as what happens after we are finished with it. Most text books are made primarily of paper, wax, ink, and plastic. The paper originates from trees logged from forests or recycled paper products, both processed with water into a slurry of paper, or pulp. The slurry is then spread out into a thin layer, dried thoroughly, and then bleached or colored and rolled into rolls for distribution. The paper might then end up at a printing shop, where the paper is cut and printed upon using various inks derived from oils. The text book cover may be a shiny, wax-covered color ink shell also produced at the shop. Finally, a plastic cover might be added to the text for shipping in a cardboard box, with some plastic protection pieces within. Once at the school, students purchase the book, use it and hopefully either continue using it, sell it back for re-use, or recycle it, so it can be made into a new, hopefully better book.

Figure 1.10 presents the basic components of LCA.[2] Beginning with resource extraction, materials are processed into the raw materials we use for product design and manufacturing. Once products are produced, they are distributed and marketed through various retail systems. We then purchase and use the products, and have a decision to make at the end of the product's utility—we can throw it away to be stored (and eventually decomposed) in a land-fill, or return the product to the material processing step, to be recycled into another product. At every step in the life-cycle, some amount of both mass and energy are required; the sum total of these comprise the environmental impact of the product under the LCA.

Ecological Footprints

Ecological footprints can be one form of data output from LCA; these measure the demands on the ecosystem by a population, product, or individual and compare with the ecosystem services provided in terms of food and energy production and removal, or digestion of waste streams. **Figure 1.11** illustrates the process of developing an ecological footprint, comparing the rate at which we consume resources and produce wastes vs. the rate at which resources are created and waste is absorbed. The term, *bio-capacity* describes the maximum population that an

Ecological Footprint Measures:

1) Resource Consumption and Waste Generation Required to Enjoy:

Energy

Seafood

Timber and Paper

Settlement

Food and Fiber

2) Compared to the Ecosystems' Ability to Generate Resources and Consume Wastes, such as:

Emissions — Built Environment — Forests — Agricultural Land — Fisheries

Figure 1.11
Ecological footprints describe demands on ecosystem services.

ecosystem can naturally support long-term in regard to resource production and waste absorption (ecosystem services). It is generally agreed that through the use of fossil fuels, humans have far exceeded the natural *bio-capacity* of Earth.

Figure 1.12 presents data from the United Nations on human development index (a measure of life-span, education, and income) vs. ecological footprints (measured in average hectare's worth of land productivity and waste absorption required per person) of different countries in the world.[3] The countries are ranked (1–177) in order of their human development index, and compared in terms of their average impacts. Most developed regions are well beyond Earth's natural bio-capacity of ~2.1 hectares per person. Many developing regions are currently below the bio-capacity in terms of ecological footprint, but are growing rapidly, which increases ecological footprints substantially. On average, individuals within the countries at the top of the ranking enjoy relatively long life expectancies, better education, and more financial wealth compared with those in the bottom. This improved life-style is often associated with larger footprints on account of the increased consumption of energy and stuff that this lifestyle affords. However, as seen by comparing top-tier countries, like Japan, Norway, and the United States, for example, a high human development ranking does not necessarily require equally higher ecological footprints.

Human Welfare and Ecological Footprints compared

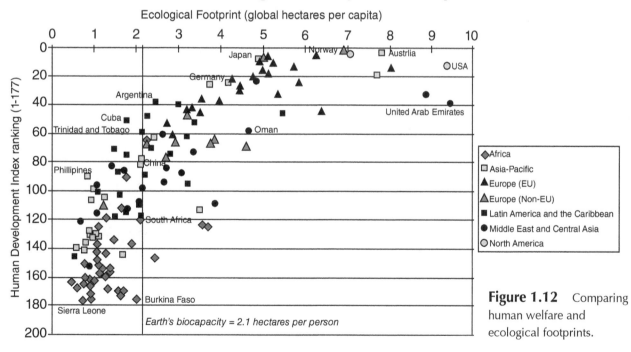

Figure 1.12 Comparing human welfare and ecological footprints.

Source: United Nations Development Programme.

References

1. http://www.mitomap.org
2. http://www.epa.gov
3. http://www.undp.org

Quiz

(Open Book—Write Answers Below Questions—Show All Work)

1. When, approximately, did modern humans (homo sapiens) first migrate out from Africa?

2. When, approximately, did the Agricultural and Industrial Revolutions occur?

3. When did the global human population reach: 1 billion_____? 2 billion_____? 3 billion_____? 6 billion_____? 7 billion_____?

4. What is the most populated continent in the world?

5. What continent is growing the fastest in terms of population?

6. What phase changes pull thermal energy from the surroundings?

7. Define HVAC and describe its major features.

8. Define LCA and name its major features.

9. Describe the term bio-capacity, and how humans exceed it.

10. Describe a simple LCA for a product you use regulary.

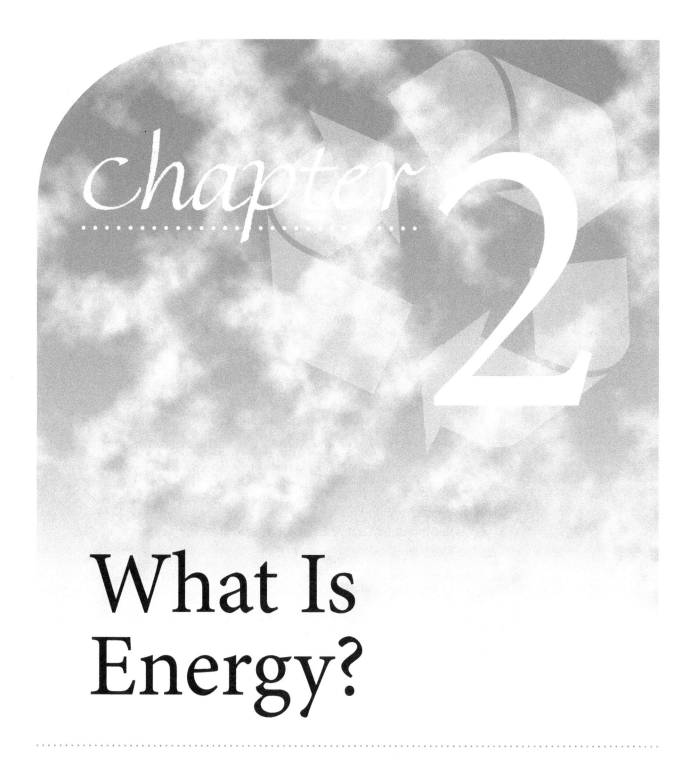

chapter 2

What Is Energy?

"It is important to realize that in physics today, we have no knowledge of what energy is."

— The late Richard Feynman, Nobel Prize Winning Physicist and Avid Bongo Player

Definitions

How is it that we humans use energy to drive cars, fly planes, and rocket into space, while our understanding of its origin and very nature remains incomplete? This question reveals the mystery of energy, which powers all life on Earth and the amenities of modern civilization. Clearly, we do understand something about energy since we have managed to manipulate it to our advantage. What we think we know about energy begins with its definition. Before proceeding, pause and think of your own definition of *energy*, and if you'd like, write it here:

_____

_____

Now, compare your definition with the six provided by the Merriam-Webster Online Dictionary[1]:

1. dynamic quality <narrative energy>
2. the capacity of acting or being active <intellectual energy>
3. a usually positive spiritual force <the energy flowing through all people>
4. vigorous exertion of power: effort <investing time and energy>
5. a fundamental entity of nature that is transferred between parts of a system in the production of physical change within the system and usually regarded as the capacity for doing work
6. usable power (as heat or electricity); also: the resources for producing such power

Of course, the meaning of any word depends on its intended use, and with the word energy, we must establish whether we are talking about the energy of a sporting event or rock concert or the electrical energy powering a home appliance. The energy at a sporting event or rock concert is a more figurative (and perhaps more common) use of the word energy, whereas the electrical energy to power an appliance is a more exact, scientific use of the word. The first four (1–4) definitions are more figurative compared to the more scientific definitions (5–6). Many science text books define energy as the *capacity to do work*, a condensed version of definition 5. Definition 6 relates energy to power, heat and electricity. For the moment, we will adopt definitions 5 and 6, while discussing other interpretations and definitions for comparison.

Unfortunately, most definitions fail to capture the universal and abstract nature of energy. In fact, energy is basic to the fabric of reality. As famous German writer, Johann Wolfgang von Goethe, once described, "Energy will do anything that can be done in the world." Energy pervades the universe and is always conserved (what goes in must come out); however, many scientists, including famed physicists, claim that we lack fundamental knowledge of *what energy really is*. Some cleverly define energy as the "mysterious everything"[2].

Energy is most often noticed when it is converted from one form to another. During the conversion, stored, or potential energy, converts into active, or kinetic energy. For example, a person running is converting stored energy from the reaction between the food they eat and the oxygen they breathe into active energy in the form of muscles contracting/expanding during the run. Another example is your cell phone, which converts stored chemical energy within its battery into active electrical energy to power the phone. In this way, both the food and oxygen for the runner and the battery for the phone have the capacity to do work or the ability to supply power. That is, until the stored energy is used up, and the runner needs to eat and breathe or the battery needs to be recharged (converting active energy back into stored energy). **Figure 2.1** presents energy conversion in action; it is happening everywhere all the time.

Whenever energy conversion takes place within physical or living systems, changes are made to the surrounding environment. For example, as the runner, or battery convert stored energy into active energy (motion or electricity), they also produce thermal energy (heat), which warms the surroundings. This permits another, basic definition of energy *the ability to change the surroundings*. In this way, energy is basically *the ability to do something*. Now we have the following definitions of energy:

- capacity to do work
- usable power
- ability to change the surroundings
- ability to do something

Figure 2.1   Energy conversions: runner converting food and oxygen (chemical energy) into motion (mechanical energy); cell phone converting stored energy in battery (chemical energy) into lights and computations (radiant and electrical energies).

Keep in mind that energy can be stored (as in food or a charged battery) or active (as in a person running or a cell phone), and is constantly converted from one form to another. For example, energy from sunlight helps grow fruits, nuts, grains, and vegetables. Energy within the food we eat is converted with the air we breathe to help us stay warm and active. Since the beginning of time, energy has been converted from one form to another in a repeated cycle, all the while making changes to the surroundings.

Energy Forms

Energy comes in various forms and is always conserved, that is its total amount never changes. This is similar to how water exists in various forms (ice, liquid, and vapor) and is conserved (never disappears). For example, when ice melts, a puddle forms. Or, if liquid water boils or evaporates, the air becomes humid. We can track every water molecule, regardless of how often it changes from one form to another.

Scientists call this the *conservation of mass*, which is basically an accounting system for mass in all of its different forms. Energy can be treated the same way; whenever one form of energy disappears, another form appears instantly. Scientists call this the *conservation of energy*, which is basically an accounting system for energy in its various forms. *The conservation of both mass and energy in their various forms are fundamental concepts for science and engineering, which is simply the application of science to solve problems.* **Figure 2.2** illustrates this essential concept. As mass or energy enter any system they must exit, otherwise they will accumulate—like water in a bathtub. Depending on the flow of mass and energy into and out of the system, and energy conversions within the system, changes are made to the surroundings.

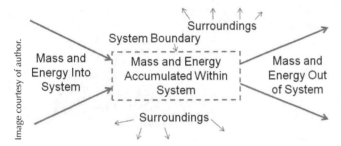

Figure 2.2 Mass and energy conservation principles.

Many textbooks focus on two main forms of energy; *potential energy* and *kinetic energy*, or as we have termed *stored energy* and *active energy*. The conversion of potential into kinetic energy lends itself to straight-forward mathematical treatment of falling objects, like predicting the trajectory of a rock falling or a skier dropping a cliff. In these cases, the potential or stored energy of the rock or skier on top of the cliff is converted into kinetic or active energy as the rock or skier falls (gravity pulling them down). Another example is a person falling down the stairs; when the person slips at the top of the stairs, potential energy is converted into kinetic energy while falling down the stairs. In all cases, the kinetic energy can also be converted into various sounds and injuries, depending upon where and how the rock or person lands.

While potential (stored) and kinetic (active) are primary energy forms, there are several sub-categories for both, which are common in everyday life. In this book, we focus on seven (7) common energy forms, which may fit into either or both potential and/or kinetic energy categories. **Table 2.1** presents the seven energy forms, along with their definitions and examples. Understanding these energy forms will help you to determine how energy is present everywhere around you. The seven energy forms can be remembered using their initials: M = Mechanical; T = Thermal; C = Chemical; E = Electrical; N = Nuclear; G = Gravitational; and R = Radiant, which is technically termed electromagnetic radiation. One method to remember these seven energy forms is to use a mnemonic device with the initials. For example, a "Montana Chemical Engineer" could be abbreviated (initialized) as, "MT-CENGR", with each of the letters representing a different energy form. It may be helpful to develop your own mnemonic device to remember the seven energy forms (the order of the letters is not important—be creative!). If you'd like, write it here:

_____

_____

Table 2.1 Common Energy Forms, Definitions, Examples and Classification

Energy Form	Definition	Examples	Potential or Kinetic
M = Mechanical	Moving things and the potential to move things	Moving car, lifting weights, stretching rubber band	Both—think of a compressed spring vs. swinging a bat
T = Thermal	Internal energy of substances—often called heat and related to their temperature	Geothermal vents, kitchen stoves, steam	Kinetic Only—thermal energy is the movement of atoms and molecules
C = Chemical	Energy within the bonds between atoms and molecules—converted through reactions into other bonds with higher or lower energies	Any fuel, food or waste plus the oxygen used to burn (or metabolize) them	Potential Only—chemical energy is stored within substances and only converted into other forms of energy through reactions
E = Electrical	Flow of electric charge	Electrical devices and appliances—everywhere	Kinetic Only—electrical energy is the flow of electric charge through a voltage
N = Nuclear	Energy stored within the nucleus of atoms	Fission = splitting atomic nuclei (nuclear power plants) Fusion = combining atomic nuclei (powering our sun)	Potential Only—nuclear energy is stored within substances and converted through nuclear fission or fusion into more useful forms
G = Gravitational	The energy of a position above some reference due to gravity pulling downward	Waterfalls, sky divers and downhill skiing/boarding/sledding	Potential Only—gravitational energy is stored and converted into mechanical energy when falling
R = Radiant*	Energy traveling in electromagnetic waves in packets called *photons*	Visible light, ultraviolet (UV) and infrared (IR), radio, x-rays	Kinetic Only—energetic waves (photons) traveling through space or materials

*More accurately termed: electromagnetic radiation.

Energy Uses (Conversion Processes)

The various energy forms are constantly being converted from one into another. Whenever this happens, other, often less-desirable, energy forms are created in the process. This means that energy conversion processes are less than perfect—in technical terms, the energy conversion *efficiency* is less than 100%, or 1.00 in fractional terms. *Efficiency* is a measure of the amount of useful energy we get out of any system divided by the amount energy we put into the system. That is,

$$Efficiency = \frac{what\ you\ get\ out}{what\ you\ put\ in}$$

This does not mean energy is lost. In fact, it is always conserved—when one form of energy is converted to another, a third (or more) less-useful form of energy is also produced. For example, a warm cell phone is a byproduct of the battery coverting an amount of chemical energy (C) into a smaller amount of electrical energy (E) with the remaining energy converted into less useful thermal energy (T). This reality is what prevents the existence of *perpetual motion* machines, or any other (non-nuclear) device that claims to provide more energy out than what it is supplied.

A common example of energy conversion efficiency is an incandescent light bulb, where electrical energy (E) is partially converted into radiant energy (R). This energy conversion process also generates thermal energy (T), warming the bulb and making it hot to the touch. In fact, only about 5% of the electrical energy is converted into light in incandescent light bulbs (R = 0.05 E), with about 95% used to warm the bulb (T = 0.95 E). In other words, these light bulbs have an energy conversion *efficiency* of only about 5%, i.e., only 5 of every 100 electrical energy units put into the light bulb were converted into radiant (light) energy out. Newer lighting technologies significantly increase this efficiency, e.g., CFLs and LEDs have efficiencies of 15 to over 60% (3 to over 20 times better).

It is important to understand energy conversion efficiency. If an energy conversion system is replaced with a more efficient one, it will use less energy for the same task, and save on energy expenses. However, the replacement costs may be higher than the original system, so calculating the long-term cost and savings of the investment can be helpful for these decisions. **Table 2.2** presents various energy conversion processes and their approximate efficiencies .[3, 4] Another common unit used to describe efficiency in gasoline or diesel-powered transportation is miles-per-gallon, or MPG. A vehicle's MPG rating describes the average distance it will travel when burning one gallon of fuel. Of course, as driving conditions change (e.g., wind, hills, surface) MPG is affected. Many vehicle MPG ratings are found at www.fueleconomy.gov.

Table 2.2 Energy Conversion Examples and Their Approximate Efficiencies [3, 4]

System	Energy Conversion	Approximate Efficiency (%)
Incandescent Light Bulbs	E → R	5
Muscles	C → M	15
Combustion Engines	C → M	25
Modern Wood Stoves	C → T	40
Small Electric Motor	E → M	70
Modern Batteries	C → E	85
Large Electric Generators	M → E	95

When multiple energy conversion systems are placed in series, their independent efficiencies are multiplied by one another. Since the efficiency number is always less that 100%, or 1.00, this multiplication can dramatically reduce the overall efficiency of series-connected energy conversion processes. For example, consider electric lighting within a remote cabin, which has a gasoline-powered electric generator outside. If the generator was 40% efficient (high estimate), and the cabin used 5% efficient incandescent light bulbs, then the overall efficiency of converting the gasoline and oxygen into lighting is 40% multiplied by 5%, or 0.40 × 0.05 = 0.02, which is only 2%. That is, only about 2% of the chemical energy (C) generated from the gasoline-air reaction was converted into radiant energy (R) to light the cabin. The remaining 98% of the chemical energy (C) was converted into low-value thermal energy (T) and noise—mechanical energy (M) during the multiple energy conversion steps:

- First, the gasoline generator inefficiently converts the chemical energy (C) in the fuel-air reaction into mechanical energy (M) spinning a shaft (C → M).
- The spinning shaft is connected to an electric generator, which is inefficiently converting mechanical energy (M) into electrical energy (E).
- The incandescent bulbs convert electrical energy (E) into radiant (light) energy (R).

Overall, from gasoline to lights, there are four energy forms and three energy conversions involved to get the desired form (C → M → E → R). This example emphasizes the importance of efficiency in our various energy conversion systems, and helps identify more direct use of energy forms (minimizing the number of conversions).

Table 2.3 presents energy conversion processes and examples of systems in which the conversions take place.[3] Beginning with energy forms in the left column, energy is converted to the forms in the top row through the processes and examples within the table's cells. While some terms may not be familiar, the general concept of energy conversion is important. Specific technologies will be presented and discussed later in the book.

Table 2.3 Energy Conversions and Example Systems[3]

Starting From ↓	Converting to:						
	Mechanical	**Thermal**	**Chemical**	**Electrical**	**Nuclear**	**Gravitational**	**Radiant**
M	Gears	Friction	Absorption	Generators	Particle Beams	Rising Object	
T	Steam Engines	Heat Exchangers	Cooking	Thermo-electrical	Fusion		Thermal Radiation
C	Muscles	Burning (combustion)	Chemical Processes	Batteries			Chemical-luminescence
E	Motors	Resistance Heaters	Charging Batteries	Power Lines			Lights
N	Nuclear Bombs	Fission and Fusion	Ionization	Nuclear Batteries	Chain Reactions		Nuclear Bombs
G	Falling Objects				Fusion		
R	Radiometers	Solar-Thermal	Photosynthesis	Solar-Electric Cells	Gamma-Neutron Reactions		

Energy Sources

Where does energy come from? Ultimately, all energy sources are products of stellar processes occurring throughout our universe's history. From these intense and complex processes, which convert tremendous amounts of mass into energy and energy into mass, come the basic building blocks for Earth and all life on it. Now, and for the next several billion years, energy from nuclear fusion reactions within our sun will continue to power the Earth. Humans have used solar energy for a variety of purposes throughout our existence. We use solar energy directly to heat our dwellings and dry our food; we use indirect solar energy, e.g., wind, hydro, and biomass, to power our transportation, mill grains, and make fire; and now, we use ancient solar energy (fossil fuels) along with other stored energy sources to power modern society. One way to remember the primary energy sources on Earth is to separate them into *new, old,* and *really-old* categories.[5]

New energy sources are all connected to the radiant energy continuously bombarding Earth from our sun (referred to as solar energy). This includes *direct* solar energy, such as the sunlight warming the land or water, or converted into electricity by solar-electric systems. New energy also includes *indirect* solar energy, such as growing crops (using photosynthesis) to feed animals and humans (all biomass), evaporating water, and producing winds to drive waves and power wind mills and turbines. These new energy sources are all considered *renewable* by virtue of their semi-continuous replacement, or at least those that do so within a reasonable time frame. At present, only about 8% of total domestic energy (heating and electricity) is supplied by new sources in the United States.

Old energy sources are products of tens to hundreds of millions years of biological and planetary processes (too long to consider them renewable). What was new energy long ago, e.g., lush tropical ecosystems (biomass), died and deposited in mud, and was sequentially buried and slowly converted into solid, liquid, and gaseous energy sources, e.g., coal, oil, and natural gas (depending on the geologic formation conditions). These energy sources are dated with fossilized records of the species living during ancient times, and are therefore termed *fossil-fuels*. Energy-dense fossil fuels have substantially facilitated modern civilization, and currently supply over 80% of both global and US energy demand.

The reserves of these old energy sources are limited—only so much past biomass was converted into fossil fuels—and therefore all fossil fuels are considered *non-renewable*. Additionally, the location of the fossil fuel reserves around Earth is geographically uneven, and many reserves are found in regions with politically-unstable governments. Moreover, the byproducts produced during their use are increasingly disrupting the Earth's natural climate regulation. However, due to the significant inertia of our fossil fuel-based infrastructure (both technological-economical and sociopolitical), the transition from fossil fuels will be challenging, and most analysts project that fossil fuels will dominate global energy supply for decades to come.

Really-old energy sources are product of billions of years of planetary processes and include nuclear and geothermal energy. Nuclear energy is most often realized by splitting heavy, unstable atoms in a process known as fission, heat from which boils water into steam to power electric generators. Nuclear fission currently supplies about 20% of electric energy demand in the United States (about 6% of total energy). Fission fuels are typically derived from uranium, which is a limited resource, thereby making nuclear fission technically non-renewable. However, there are various nuclear fuel cycles which utilize other, more Earth-abundant elements, and if safely developed along with fuel reprocessing, nuclear fission fuel reserves could last much longer. Geothermal energy results from planetary and nuclear fission processes deep within the Earth and is generally considered renewable because of the enormous magnitude of the energy available. The United States currently produces less than 1% of total energy from geothermal sources.

Figure 2.3 presents the different primary energy sources and their interconnections to typical energy forms. Solar energy drives most of the processes, whether from the renewable flux of new solar energy, or stored geologically from old solar energy (fossil fuels). The Earth's core, crust, and oceans also provide energy through geological and tidal processes.

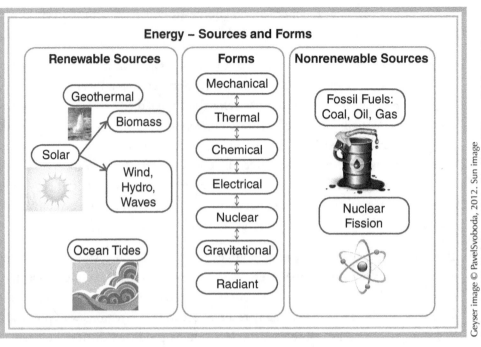

Figure 2.3 Energy sources (renewable and non-renewable) and energy forms.

Figure 2.4 illustrates the breakdown of energy sources and demands within the United States in 2011, using energy units of quadrillion British thermal units, or BTUs and in percentages. The energy sources are shown on the left, with energy demand sectors on the right. From this, it is apparent that the fossil fuels dominate the US energy portfolio. Additionally, it is clear that the US transportation sector and the electric power sector are heavily dependent on petroleum, and coal, respectively.

Figure 2.4 2011 US primary energy flow by source and demand sector.

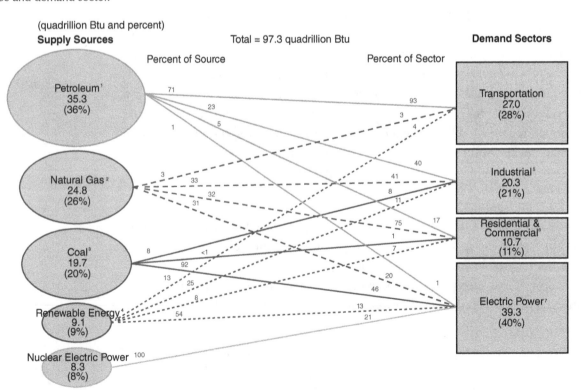

Source: US-DOE-EIA. http://www.eia.gov/totalenergy/data/annual/pecss_diagram.cfm

Extraction of both renewable and non-renewable energy sources and conversion into useful energy forms is a critical foundation of modern society. However, before delving into energy conversion technologies it is important to first learn common terms and metrics. We can then directly compare different energy sources and conversion technologies in terms of their energy densities (how much energy stored per unit mass or volume) and the efficiencies of various energy conversion processes. These comparisons are imperative for effective decision-making personally in the household and in society at-large regarding which type of energy sources and technologies to employ, given resource and economic constraints.

Power, Work, Heat and Metrics

"For those who want some proof that physicists are human, the proof is in the idiocy of all the different units which they use for measuring energy."

— Richard Feynman, The Character of Physical Law (1967)

In 1960, the International System of Units (SI) was adopted to standardize units of measure (metrics) in science. There are seven (7) basic SI units, from which the more complicated units are derived. These seven units (**Table 2.4**) are used to describe:

1. Length, measured in meters (m)—1.0 m = 3.3 feet (about an arm's length).
2. Mass, measured in kilograms (kg)—1.0 kg = 2.2 pounds (a small sack of spuds).
3. Time, measured in seconds (s)—1.0 second = 1/60 minute (about one resting heartbeat).
4. Temperature, measured in Kelvin (K)—273 K = 0°C = 32°F (water freezing into ice).
5. Electric current, measured in Amperes (A)—1.0 A = 1.0 Coulomb of charge per second (about that required for a small fan).
6. Amount of substance, measured in mole (mol)—1.0 mol = 6.02×10^{23} (about the number of carbon atoms in a lump of coal).
7. Luminous intensity, measure in candela (cd)—1.0 cd = 1.0 lumens per steradian (about the light put out from a tiny LED).

While these are standard units for measurement, many different measurement systems evolved throughout history. This has left us a legacy of numerous different units, such as feet and pounds vs. meters and kilograms. The common units used in the United States are related to the SI units later in this chapter. Converting among the basic units of measurement is now facilitated by unit conversion software on many websites and hand-held devices, and in some cases, it is extremely important to get correct! For example, in 1999 a mistake with unit conversion caused NASA's $125 million Mars Climate Orbiter to fly too close to the Martian atmosphere and it disintegrated.

Table 2.4 Basic SI Units, Names and Symbols

SI Unit of:	Name:	Symbol:	English Unit:
Length	Meter	m	Foot (ft) 1 ft ≈ 0.3048 m
Mass	Kilogram	kg	Pound (lb) 1 lb ≈ 0.4536 kg
Time	Second	s	-
Electrical Current	Ampere	A	-
Temperature	Kelvin	K	Fahrenheit (°F) 32 °F = 273.15 K
Amount of Substance	Mole	mol	-
Luminescence	Candela	cd	-

We know that energy can be defined as the *capacity to do work,* or *useful power (as heat or electricity).* These definitions necessitate additional definitions of *power, work* and *heat. Power* is the rate of energy use (how much energy expended per unit time). This makes energy the amount of power multiplied by the time it's applied. *Work* is simply energy in either mechanical or electrical form. *Heat* is the transfer of thermal energy resulting from temperature differences (think of heating water).

Mechanical work is defined as force multiplied by distance, e.g., lifting or turning something, or pressure multiplied by volume, e.g., piston-cylinder. The basic SI unit for mechanical work is the Joule (J), which is a force multiplied by a distance. The standard unit of force, the Newton (N), is mass multiplied by acceleration (kg-m/s^2). When multiplied by the standard unit of length, the meter (m), a Newton-meter, or 1 Joule, results. Therefore, a Joule is equal to a Newton-meter (1 J = 1 N-m = 1 kg-m^2/s^2—about the about the energy required to lift a small bag of potatoes off the floor onto a table). The amount of mechanical work performed over a specific time (mechanical power) is measured in terms of a Watt (W), which is defined as one Joule per second (1 W = 1 J/s = 1 kg-m^2/s^3—about the power exerted in lifting the bag of potatoes in 1 second). Common units and descriptions for mechanical force, work, and power are presented in **Table 2.5**.

Electrical power uses the same unit as mechanical power, the watt (W). Electrical power comes from the flow of electric current, measured in Amperes (A), driven by a pressure, or potential, measured in volts (V). A watt of electrical power is defined as one ampere (A) of electric current through one volt (V) of electric potential (1 W = 1 V × 1 A = about the power required for an LED keychain light). Electrical work (electric energy) is then a watt of electrical power multiplied by time that it is applied, measured in watt-seconds (W-s) which is the same as one Joule (J). Since we use so much electricity, we normally measure it using kilowatt-hours (kW-hrs),

Table 2.5 Descriptions and Units for Mechanical Force, Work, and Power

Thing	What is it?	SI Unit	Common Unit Name	English Unit
Force	Mass × Acceleration	$kg\text{-}m/s^2$	Newton (N)	Pound-force (lbf) 1 lbf ≈ 4.448 N
Work	Force × Distance (Energy)	$kg\text{-}m^2/s^2$	Joule (J) = N-m	Foot pound-force (ft-lbf) 1 ft-lb ≈ 1.356 J
Power	Energy / Time	$kg\text{-}m^2/s^3$	Watt (W) = J/s	Horsepower (HP) 1 HP ≈ 745.7 W

Table 2.6 Descriptions and Units for Electrical Power and Work

Thing	Description	SI Unit	Common Unit
Electrical Power	Voltage × Current	$kg\text{-}m^2/s^3$	Watt (W)
Electrical Work	Power × Time (Energy)	$kg\text{-}m^2/s^2$	Joule (J) = W-s (Often kW-hr)

which is 1,000 W (1 kW) of power applied for 1 hour, or about the energy required to dry a small load of damp towels in an electric dryer. **Table 2.6** presents useful information for electrical work and power.

Coincidently, heat also uses the same units as mechanical and electrical work. In fact, since work and heat are both energy forms, this would include work as mechanical or electrical energy (M or E) and heat as thermal energy (T). When you think of heat, remember temperature differences (thermal energy naturally flows from higher to lower temperature via *heat transfer*). Heat is often associated with the concept of heating things. For example, when we heat up water on an electric-powered stove, we are converting electrical energy into high temperature thermal energy (heat), which is transferred into the lower temperature water making its temperature increase. With some limits, heat can also be converted to work, and vice-versa. This equivalency between heat and work is the foundation of the science of *thermodynamics*, which describes natural driving forces (toward equilibrium) as well as the operation of many modern energy conversion devices.

Heat is transferred in one, or a combination, of three ways: conduction, convection, and radiation. Conductive heat transfer between objects occurs when the objects at different temperatures are in physical contact (touching a cold flag pole). Convective heat transfer happens, for example, when warm air is physically blown out of a furnace with a fan (moving hot air). Radiant heat transfer occurs when energy is transferred via photons (electromagnetic waves), typically in infrared (IR) radiation. The three forms of heat transfer are illustrated in **Figure 2.5**.

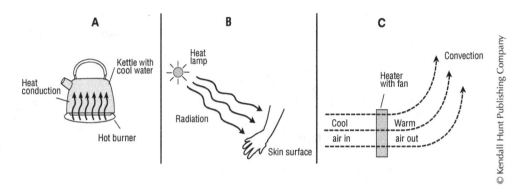

Figure 2.5 Heat transfer by: a) conduction; b) radiation; and c) convection.

© Kendall Hunt Publishing Company

Three major temperature scales, Fahrenheit (°F), Celsius (°C) and Kelvin (K), ranging from absolute zero (0 Kelvin) to the normal boiling point of water, are presented in **Figure 2.6**. The amount of thermal energy (heat) required to increase the temperature of 1 kg of liquid water by 1 °C is called a *kilocalorie* (kcal). This is also known as the *dietary calorie* (Calorie = kcal) and is used to measure the energy content of foods (when metabolized with oxygen from breathing air). For example, a recommended daily diet in the United States is approximately 2,000 Calories (2,000 kcal), and there are often over 500 Calories (500 kcal) in fast-food hamburgers. In the United States, the British thermal unit (BTU) is used to describe the amount of heat required to increase the temperature of 1 lb of liquid water by 1 °F. In both thermal energy (heat) units, the thermal power is defined as the amount of heat over unit time (heat/time). **Table 2.7** presents useful information for thermal energy (heat) and thermal power.

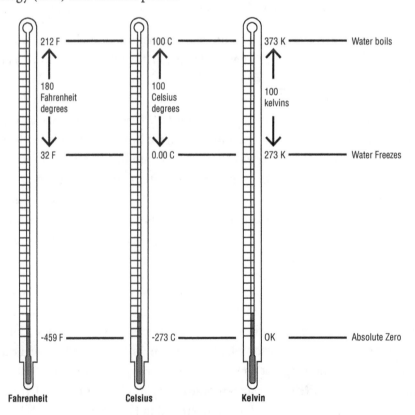

Figure 2.6
Temperature scales.

Source: NASA

Table 2.7 Heat and Thermal Power Descriptions and Units

Thing	Description	Common Unit	English Unit
Thermal Energy (Heat)	Energy required to increase temperature of substance	1 kilocalorie (kcal) increases 1 kg water by 1 °C	1 British thermal unit (BTU) increases 1 lb water by 1 °F
Thermal Power	Heat / Time	kcal/s	BTU/min

Common units for energy and power and their conversions are presented in **Tables 2.8** and **2.9**. To convert from one unit of energy or power to another, begin with the unit in the top row (Tables 2.8/2.9) and move down that column until intersecting the row with the desirable unit. For example, if we were to determine how many BTUs are in 10 kW-hrs (both energy units), we begin with the left-most column in Table 2.8, and move downward to the row related to kW-hr row. Beginning with 10 kW-hr, we move right on the row to the intersection with BTU column, where the number 3,412 is located. This means that for every one (1) kW-hr, there are 3,412 BTUs. Therefore, in 10 kW-hrs, we would have:

$$10 \ kWhrs \times \frac{3412 \ BTU}{1 \ kWhrs} = 34{,}120 \ BTU$$

Similarly, if we had an engine with a power capacity of 100 horsepower (HP) and wanted to know how many kilowatts (kW) this was worth, we would use values in Table 2.9. Beginning with the HP row, we proceed right to intersect with the kW column and find the number 0.7457. This means that every one (1) horsepower is equivalent to 0.7457 kW. Therefore, our 100 HP is:

$$100 \ HP \times \frac{0.7457 \ kW}{1 \ HP} = 74.57 \ kW$$

Table 2.8 Common Energy Units and Approximate Conversions

Starting from units of:	Multiply by the conversion values to get below units				
J	1	0.001	9.478×10^{-4}	2.388×10^{-4}	2.778×10^{-7}
kJ	1,000	1	0.9478	0.2388	2.778×10^{-4}
BTU	1,055	1.055	1	0.252	2.931×10^{-4}
kcal	4,186	4.186	3.968	1	1.163×10^{-3}
kW-hr	3,600,000	3,600	3,412	859.8	1
	Joule (J = W–s)	KiloJoule (kJ =–kW–s)	British Thermal Unit (BTU)	Kilocalorie (kcal)	Kilowatt-Hour (kW–hr)

Table 2.9 Common Power Units and Approximate Conversions

Starting from units of:	Multiply by the conversion values to get below units				
W	1	0.001	0.0569	1.341×10^{-4}	2.388×10^{-4}
kW	1,000	1	56.869	1.341	2.388
BTU/min	17.58	0.0176	1	0.236	4.2×10^{-3}
HP	745.7	0.7457	42.41	1	5.615
kcal/s	4,187	4.187	238.1	5.615	1
	Watt (W)	Kilowatt (kW)	BTU per minute (BTU/min)	Horsepower (HP)	Kilocalorie per second (kcal/s)

In science, we often encounter very large and very small numbers as a result of studying different length scales, like the distances in between planets vs. the thickness of a single strand of hair. To make it easier, scientific notation is often used to describe these very large and very small numbers by using multiples of 10. Every multiple of 10 is called an *order of magnitude*. For example, since the distance between Earth and Mars is about two hundred billion meters, or 200,000,000,000 m, it is easier to write this as 200×10^9 m; simply by replacing the nine zeros following 200 with a multiple of ten (10) to the power of nine. The thickness of a human hair is about one hundred micrometers (microns—one millionth of a meter), or 0.0001 m, and it is often easier to write this as 100×10^{-6} m, or one hundred times ten to the power of negative six. Comparing the distance between Earth and Mars ($\sim 200 \times 10^9$ m) and the thickness of a human hair ($\sim 100 \times 10^{-6}$ m), we find that these two distances are fifteen (15) orders or magnitude different in length (9−(-6))!

Common multiples and submultiples using scientific notation, along with prefixes, abbreviations, and English names are provided in **Table 2.10**. It is important to familiarize yourself with the concept of describing very large and very small numbers so that you can better understand the magnitude of interactions among various physical and living systems on Earth and elsewhere in the universe.

It is often useful to have a rough comparison of common fuels (energy sources) and their energy densities (how much energy released in their combustion per mass or volume). **Table 2.11** compares the energy density of various fuels separated into foods, fossil fuels, and biomass.[6] Keep in mind that, as we convert these energy sources into useable forms, the process is often substantially less than 100% efficient. Additionally, it is important to recall that the biomass (including foods) and fossil fuels are both ultimately the result of photosynthesis; however, the fossil fuels are non-renewable, whereas the biomass resources (for the most-part) are renewable.

Table 2.10 Common Unit Multiples and Submultiples with Prefixes, Abbreviations, and English Names

Scientific Notation	Prefix	Abbreviation	English Name
10^{15}	Peta	P	Quadrillion
10^{12}	Tera	T	Trillion
10^{9}	Giga	G	Billion
10^{6}	Mega	M	Million
10^{3}	Kilo	K	Thousand
10^{-3}	Milli	M	Thousandth
10^{-6}	Micro	μ	Millionth
10^{-9}	Nano	N	Billionth
10^{-12}	Pico	P	Trillionth
10^{-15}	femto	F	Quadrillionth

Table 2.11 Approximate Energy Density of Various Fuels (Energy Sources) by Category and Type[6]

Category	Type	Approximate Energy Density (MJ/kg = 10^6J/kg)
Food (refined biomass)	Carbohydrates	17
	Fats	39
	Proteins	17
Fossil Fuels	Coal	30
	Oil (Crude)	42
	Natural Gas	55
Biomass	Dry Straw (wheat or alfalfa)	18
	Dry Dung	12
	Dry Wood	20

References

1. http://www.merriam-webster.com/dictionary/energy

2. http://www.ftexploring.com/me/everything.html

3. Energy: Beginners Guide—Vaclav Smil—Oxford, 2006.

4. Tester, Jefferson, et. al. Sustainable Energy: Choosing Among Options. :MIT Press, 2005.

5. Wright, Judith, and James Conca. 2007. The GeoPolitics of Energy: Achieving a Just and Sustainable Energy Distribution. BookSurge Publishing.

6. http://physics.info/energy-chemical/

Quiz and Problems
(Open Book– Write Answers Below Questions—Show All Work)

Quiz
Answers in Spaces and Show All Work:

1. What are the seven (7) common forms of energy?

2. Describe an energy conversion process involving three (3) different energy forms.

3. Provide an example of each energy source; new, old, and really-old.

4. Describe the three (3) ways in which heat is transferred?

Problems
Answers in Spaces and Show All Work:

1. Utilizing data in Table 2.8, convert the following energies.

    1,000 kW-hrs (an average monthly electrical home use) = _____BTU

    2,000 kcal (an average daily diet) = _____BTU

    10,000 BTU (an average hot shower) = _____kcal

    10 kJ (lifting a potatoe up 4 ft) = _____kcal

2. Assume that one (1) ton of coal equivalent (TCE) equals 30 million BTU (1 TCE = 30×10^6 BTU). If a power plant is 30% efficient in converting coal into electric energy, and it is fed coal at a rate of 10 TCE per minute, how much electric power does the plant produce (in kW)?—How much thermal energy is lost each hour (in BTU)?

3. Consider the following: a power plant is 40% efficient in converting chemical energy (say by methane combustion) into electrical energy; electrical transmission lines are 85% efficient in transporting electrical energy to your home; and the light-bulbs in your home are 5% efficient in converting electrical energy into radiant (light) energy. What is the overall efficiency of converting chemical energy (methane) into radiant energy (light) for your home? Please sketch this process briefly describing <u>all</u> energy conversion processes. Also describe ways to increase this efficiency.

4. Knowing the seven common energy forms, match the energy conversions with the descriptions:

   a) $R > T > E$ b) $C > M > E$ c) $R > C > T$ d) $G > M > E$ e) $N > T > M$

   _____Solar energy is used to grow biomass (e.g., wood), which is then burned for space heating.

   _____Solar energy is focused on a water tube to produce steam, which turns an electric generator.

   _____Nuclear reactions boil water into steam, which turns a crank.

   _____Water flows downhill through a turbine, which turns an electric generator.

   _____A student metabolizes food to peddle bike connected to electric generator.

chapter 3

Earth System Science

"We travel together, passengers on a little spaceship,
dependent upon its vulnerable reserves of air and soil,
all committed for our safety to its security and peace;
preserved from annihilation only by the care, the work,
and, I will say, the love we give our fragile craft."

— Adlai Stevenson, Former US Ambassador to
the United Nations (1964)

The Earth System

Fueled by the sun, our planet Earth travels through space and supports life, much like a spaceship. That perspective can be difficult to grasp, but it accurately describes the interconnected nature and vulnerability of Earth's life-support systems. With continuous energy input from our sun and Earth's geology, complex interactions among diverse, non-living systems originated and have evolved life (as we know it) on Earth for more than three billion years. Energy flows throughout Earth's living and non-living systems determine their interactions and evolution throughout time. Earth's living systems have intimately interacted with its non-living systems to sustain environmental conditions favorable for life on Earth. Within the past two centuries, human activities have made profound changes to living and non-living systems on Earth, affecting energy flows and disrupting long-term (>10,000 year) stability of environmental conditions (climate). Understanding these disruptions will help inform strategies to minimize the impact of human activities and adapt to abnormal environmental changes. We first need to know how the Earth spaceship works, and if/how it can accommodate its most demanding passengers ever (humans).

Scientists frequently use the term *Earth System* to describe the complex, life-supporting interactions among Earth's major living and non-living systems. Referred to as *spheres* to indicate their global extent, four (4) major systems interact in intricate ways to produce the environmental conditions found throughout Earth. **Table 3.1** describes Earth's four major systems, presented in approximate order of their evolution.

Table 3.1 Earth's Major Systems and Descriptions

Earth System Component	Description
Lithosphere	Earth's crust covering its core (both ocean floor and all continents)
Hydrosphere	Earth's water in all forms (solid/liquid/vapor)
Atmosphere	Thin layer of gas (air) covering Earth's surface
Biosphere	Life on Earth in all its variable forms

Interactions among Earth's major systems and their subsystems happen at drastically variable time scales (seconds, days, centuries, millions of years) and distance scales (microscopic, local, regional, global). Meteorological metrics, e.g., temperature, atmospheric pressure, cloud cover, sun position, precipitation, air-quality, etc. over a period of time are often used to describe the environmental conditions of a region. Changes in environmental conditions on an hour-to-hour, or day-to-day basis describe the region's *weather*. If we consider average weather over seasons, years, decades, or centuries we can describe the region's *climate. Weather and climate are distinct and different—they measure the same things, but over drastically different time scales.* Weather influences our choice of clothing, e.g., swimming suit or parka, whereas climate influences our choice of crops, e.g., tropical fruits or potatoes. Climate conditions also affect how living systems evolve and migrate. Although occurring at different time scales, both climate and weather are products of complex interactions among Earth's four major system components. We now review these systems in approximate order of evolution on Earth's surface. (Table 3.1).

Earth's Lithosphere

About four and a half billion years (4.5×10^9 years) ago, planet Earth formed during a process known as *accretion*, essentially gravity pulling together adjacent mass (from within our solar system) into the enormous molten mass that was the hot young Earth. After about a billion years passed and Earth cooled, it formed into a spherical shape with distinctive layers, as shown in **Figure 3.1**. Earth's inner core is a hot and dense iron- and nickel-based solid sphere, held in shape by extreme pressure from the flowing liquid iron and sulfur core surrounding it. The movement of the iron (a magnetic material) in Earth's core creates a convective dynamo, which generates a protective magnetic field, extending far beyond Earth's atmosphere and shielding Earth's surface (and life on it) from deadly radiation from the sun and elsewhere in space. Above Earth's core are the two mantles, which are flowing layers of various molten metals and their oxides, sulfides, and salts.

Reprinted with permission from the Minerals Education Coalition. From John Christensen-Teri Christensen, *Global Science: Earth/ Environmental Systems Science, 7th Edition* (Dubuque, IA: Kendall Hunt Publishing Company, © 2009).

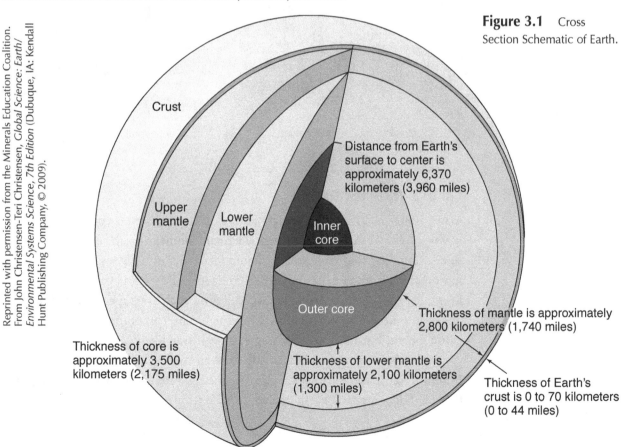

Figure 3.1 Cross Section Schematic of Earth.

Crust

Upper mantle

Lower mantle

Inner core

Distance from Earth's surface to center is approximately 6,370 kilometers (3,960 miles)

Outer core

Thickness of mantle is approximately 2,800 kilometers (1,740 miles)

Thickness of core is approximately 3,500 kilometers (2,175 miles)

Thickness of lower mantle is approximately 2,100 kilometers (1,300 miles)

Thickness of Earth's crust is 0 to 70 kilometers (0 to 44 miles)

During the cooling process, Earth's outermost layer formed into a thin rocky crust—its relative thickness like that of tinfoil covering a basketball. As the crust formed over Earth, it separated into about a dozen individual pieces, called *tectonic plates*. The plates continue to form and push into and under one another, building continents and mountains and powering earthquakes, geysers, and volcanoes, as illustrated in **Figure 3.2**. These tectonic processes created the Earth's surface topology (ocean floors, continents, mountains, ridges, valleys, etc.) and also ejected tremendous amounts of volatile gases and vapors above the crust, spawning the early atmosphere. Earth's rocky crust, or *lithosphere*, contains many raw materials (metals, minerals, fossil fuels, etc.) that we use to support and power modern civilization.

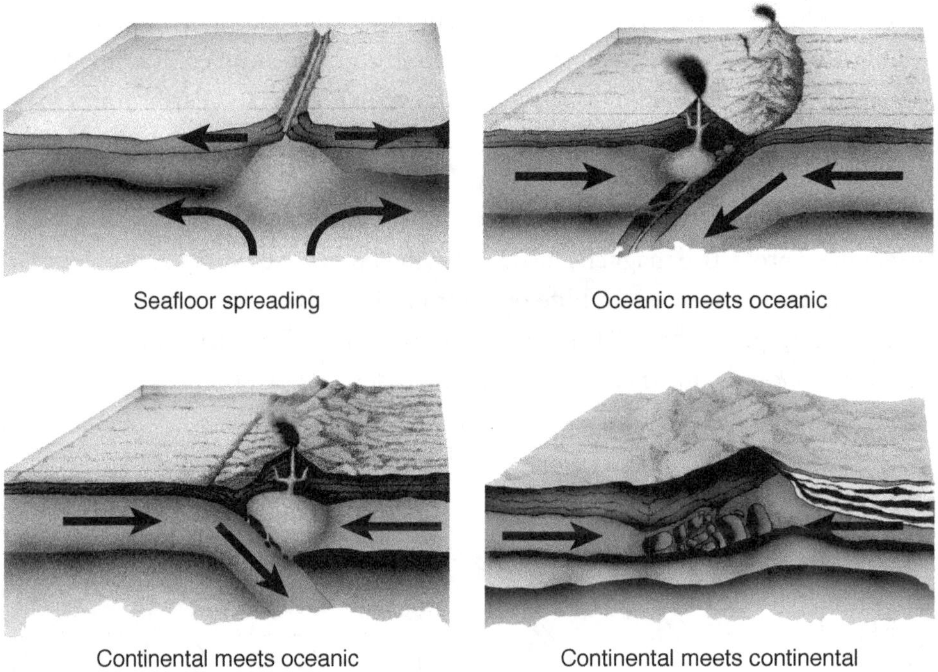

Seafloor spreading

Oceanic meets oceanic

Continental meets oceanic

Continental meets continental

Figure 3.2 Plate Tectonics on Earth.

Reprinted with permission from the Minerals Education Coalition. From John Christensen–Teri Christensen, *Global Science: Earth/Environmental Systems Science, 7th Edition* (Dubuque, IA: Kendall Hunt Publishing Company, © 2009).

Earth's Hydrosphere

Water (H_2O) is a unique chemical compound, and is often considered the universal solvent. Water is almost everywhere on and near Earth's surface, it will dissolve everything over time, and its liquid form supports life (as we know it). Like most materials, water exists in solid (ice or snow), liquid and gas (vapor or steam) forms, depending on environmental conditions, such as temperature and pressure. The origins of water (and many other elements and compounds) on Earth are not entirely clear, but most scientists agree that it was likely a combination of "outgassing" volatile compounds from the molten mantle, in addition to deposition from asteroids, and various chemical processes interacting with radiation from the sun. Soon after the Earth cooled, formed its crust, and developed a primitive atmosphere, temperature and pressure conditions were favorable to form liquid water and Earth's primordial oceans, which covered all but a few volcanic islands (before tectonic plate activity formed the continents).

As continents formed, and Earth's surface topology began taking shape, the modern version of the water cycle began, which transported water in its various forms throughout the continuously evolving lithosphere, atmosphere, and emerging biosphere. Illustrated in **Figure 3.3**, Earth's water cycle is driven by heat (thermal and radiant energy) from the sun, which vaporizes water from its liquid and solid forms through evaporation, transpiration, and respiration processes. The water is then transported into and throughout the atmosphere in the form of invisible vapor and denser clouds. As the clouds cool, rain or snow forms and falls to the ground, where it builds up, melts or pools, and is transported down to the oceans via surface or underground routes. The water cycle is one example of a *biogeochemical cycle*—one of many life-supporting cycles within the Earth System.

Source: USGS: http://ga.water.usgs.gov/edu/watercycleprint.html

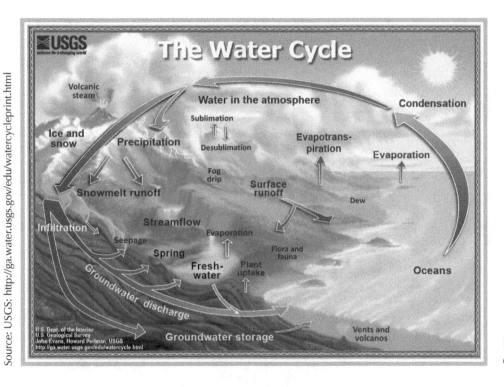

Figure 3.3 The Water Cycle on Earth.

Earth's Atmosphere

Earth's atmosphere evolved through complex interactions with its lithosphere, (ejecting volatile compounds from volcanoes), hydrosphere (gasses dissolving in oceans or water vapor in air), and biosphere (respiration from plants and animals). Earth's atmosphere functions to filter out ultraviolet (UV) radiation from the sun and maintain surface temperatures on Earth favorable for liquid water (on average). Earth's present day atmosphere consists of several layers, as illustrated in **Figure 3.4**, which also defines the boundaries between these layers. While each layer has its own distinct composition and function, the two nearest layers,

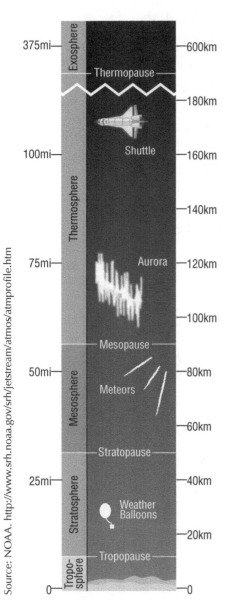

Figure 3.4 Schematic of Earth's Atmospheric Layers with Boundary Layers Defined

comprising over 90% of the atmosphere's mass, are the *stratosphere* and lower *troposphere*, which is the air we live in. Like the very outer skin on an onion, the layered atmosphere forms an extremely thin covering over Earth's surface.

The pressure of the atmosphere is equivalent to the average amount of force the atmosphere exerts against a given area. At sea-level on Earth, the air pressure is about 1 *atmosphere* (atm = 14.7 pounds per square inch, or psi), a common unit for measuring pressure. As we climb above sea level, the air pressure decreases substantially. At about 3.4 miles above sea level (~18,000 ft. or ~5,500 m.), the pressure is about half its value at sea level, or about 0.5 atm. Going approximately 3.4 miles further up to ~6.8 miles above sea level where commercial airliners cruse (~36,000 ft. or ~11,000 m.), the pressure is again halved, to about 0.25 atm. This approximate elevation also forms the boundary between the troposphere and stratosphere. The stratosphere then continues upwards for more than 30 miles (158,400 ft. or 48,280 m.), before reaching the wispy outer atmosphere layers (mesosphere, thermosphere and exosphere) interfacing with outer space.

Figure 3.5 illustrates the gradients of temperature within the atmospheric layers. Like pressure, temperatures also change dramatically within troposphere and stratosphere, however, they do not continually

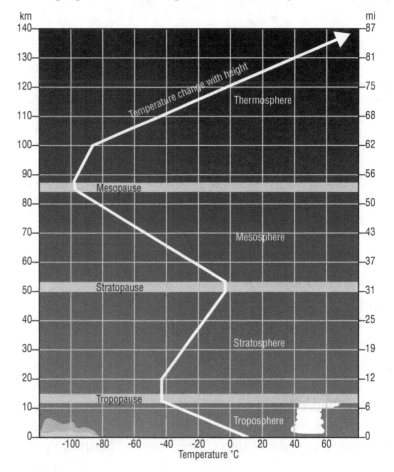

Figure 3.5
Temperature Profile in Earth's Atmosphere.

decrease with elevation. At the surface (near sea level), on average, the Earth's temperature is about 15°C (~60°F). As we climb above sea level, the temperature drops. Where commercial airliners fly, the temperatures are a frigid -40°C. Temperatures remain low and stable as we climb higher into the interface with the stratosphere before increasing again within the stratosphere. This is an important observation in understanding Earth's atmosphere and climate. The troposphere effectively traps higher temperatures close to the Earth's surface, making liquid water (and physical conditions for life) favorable.

Two of the functions of Earth's atmosphere are to block UV radiation and maintain favorable surface temperatures. The first function is primarily provided by a thin layer of ozone (O_3) gas within the stratosphere, which absorbs UV radiation before it gets to the Earth's surface. The second function occurs in the troposphere, where a small fraction (<4%) of the gases in the air trap heat by absorbing infrared (IR) radiation, which is re-radiating from Earth after it is warmed by the sun. This process, known as the *greenhouse effect* is illustrated in **Figure 3.6**, and it plays a critical role in Earth's climate dynamics. Carbon dioxide (CO_2) is a major greenhouse gas, and its concentration substantially regulates the average surface temperature on Earth, which governs the amount of water vapor in the troposphere (another strong, but more transient greenhouse gas).

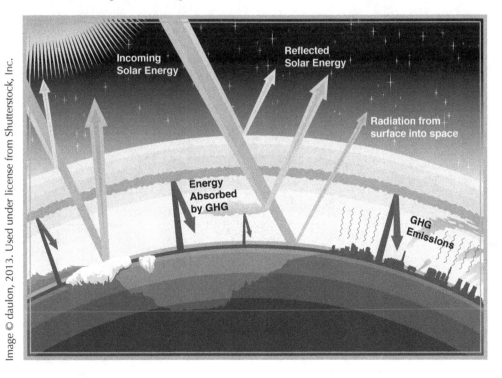

Figure 3.6 The Greenhouse Gas (GHG) Effect on Earth.

Earth's Biosphere

Simple forms of life on Earth are thought to have first developed soon after the oceans formed over three billion years ago. One model speculates that volatile gases from volcanoes, combining with energy from lighting formed basic organic chemicals, which, when dissolved in water, react to form basic amino acids—the building blocks of life. From its rudimentary form of single-cell species of archae and eubacteria, life evolved slowly over billions of years into five distinctive kingdoms, as shown in **Figure 3.7**. These kingdoms (plants, animals, fungi, procists, and bacteria) differ in the way they process energy within their environments. Within each of the kingdoms, life is further classified into phylum, class, order, family, genus, and finally species, all related to the ecological niche and genetic code that define them. Simple bacteria can parasitically use the energy within a human or other animal host to survive. These microscopic bacteria species interact with the host animal's sub-systems in *symbiotic* ways to promote the health of the animal, and therefore the bacteria's environment. Symbiotic relationships form the foundation for very successful living systems. In fact, human bodies have an order of magnitude more bacterial cells than human cells. These bacterial cells aide humans in everything from digestion to breast feeding.

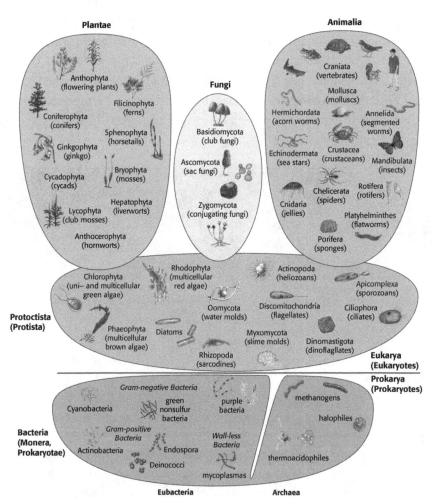

From *BSCS Biology: A Human Approach*, 3/E by BSCS. Copyright © 2006 by BSCS. Reprinted by permission of Kendall Hunt Publishing Company.

Figure 3.7   Life on Earth Classified into Kingdoms.

Plants form a basic foundation for more complex life through a process known as *photosynthesis*. Using direct solar energy, photosynthetic species convert carbon dioxide (CO_2) and water (H_2O) into oxygen (O_2) and carbohydrates (CH_2O), as shown in Equation 3.1.

$$CO_2 + H_2O + Energy = CH_2O + O_2$$

Equation 3.1

This process is illustrated in **Figure 3.8**. Carbohydrates form the basic building blocks for *biomass*, and through biological processes (governed by genetic codes), carbohydrates are converted into the numerous complex biochemicals that comprise life on Earth. Since photosynthesis consumes carbon dioxide and emits oxygen, it also changes the Earth's atmospheric composition.

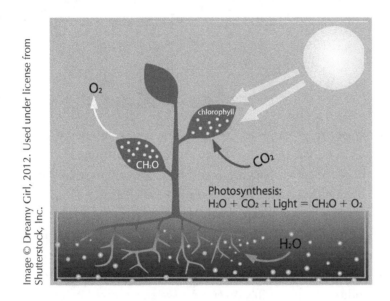

Image © Dreamy Girl, 2012. Used under license from Shutterstock, Inc.

Figure 3.8 Basics of Photosynthesis.

Bacteria (monera) and algae (protista) coexist with animals, plants, and fungi in complex communities that have co-evolved throughout time. These living communities interact with all Earth system components in extremely diverse and complex *ecosystems*. In concert with non-living systems, ecosystems provide various services in terms of basic food production and waste absorption for its inhabitants. Ecosystem services are a very important feature in Earth's climate system and human society.

Over millions of years, ecosystems evolved diverse and complex energy flows in the form of food webs, with *trophic levels* describing where species within the community rank in the energy flow. A generic North-American ecosystem is presented in **Figure 3.9**. The trophic levels include the producers (all photosynthetic species on land or in sea), the primary consumers (small animals, birds, and fish eating the producers), the secondary and tertiary consumers (larger animals, birds, and fish that eat smaller ones and the producers), and the decomposers, or micro-consumers (fungi and bacteria that eat the dead producers and consumers and produce nutrients for the producers). The number and complexity of different interacting species in ecosystems is called *biodiversity*, which also provides a metric to assess the health and resilience of ecosystems.

Ecosystems can have numerous levels of consumers; however, the effective energy transfer efficiency between each trophic level is only about 10% (from the original

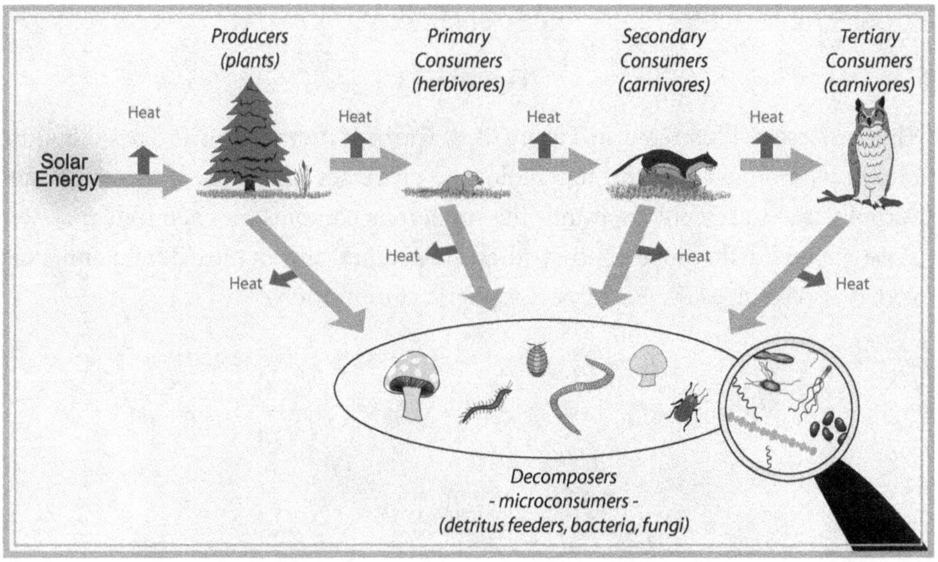

Reprinted with permission from the Minerals Education Coalition. From John Christensen-Teri Christensen, *Global Science: Earth/Environmental Systems Science, 7th Edition* (Dubuque, IA: Kendall Hunt Publishing Company, © 2009).

Figure 3.9 Trophic Levels in an Ecosystem.

conversion of solar energy in producers), which limits the number of levels. For example, if the tertiary consumers only eat the secondary consumers, they are living from only 0.1% of the ecosystem's original solar energy input (that was converted into biomass by photosynthesis, which is only ~5% efficient). From 100% of the solar energy converted into biomass by the producers, the energy conversion efficiency to primary consumers is about 10%, that of the conversion from primary to secondary consumers is about 10%, and the final conversion to tertiary consumers is about 10% (100% × 10% × 10% × 10% = 0.1%). Depending upon the long-term climate conditions, incredibly diverse ecosystems evolved in different regions and at different rates on Earth for over three billion years. Millions, if not billions, of living species have come and gone throughout this time, constantly evolving to adapt to environmental conditions and carry out life's basics; survival, procreation, and endurance.

Climate Dynamics

Earth's climate is a product of complex interactions among its major systems and subsystems. Throughout Earth's climate history, changes have been happening as a result of various internal and external *forcing factors* and *responses* through interacting components. An illustration of this process is provided in **Figure 3.10**. Historically, three (3) primary forcing factors have acted upon Earth's climate. These are changes in:

1) plate tectonics;
2) Earth's orbit; and,
3) solar intensity.

Each of these forcing factors can have significant impact on the interactions within the Earth System. Plate tectonics shift continents, build mountains, and spawn volcanoes, all of which change the climate. Changes in Earth's orbit bring it closer and

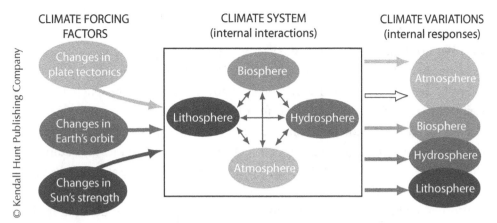

CLIMATE FORCING FACTORS

CLIMATE SYSTEM (internal interactions)

CLIMATE VARIATIONS (internal responses)

Figure 3.10 Climate Dynamics, Forcing Factors, System Interactions and Responses.

further from the sun, thus changing solar energy inputs. And, changes in the sun's intensity have a direct relationship with changes in Earth's climate. All of these changes affect the energy flows within the Earth System at different time scales, and directly influence both short and long-term environmental conditions (weather and climate).

When these climate forcing factors change, Earth's four major systems react and interact, resulting in fast or slow responses. The continuous energy from the sun drives various life-supporting *biogeochemical cycles* on Earth, such as the water cycle presented in Figure 3.3, and the carbon cycle, which is illustrated in **Figure 3.11**.

Figure 3.11 Global Carbon Cycle.

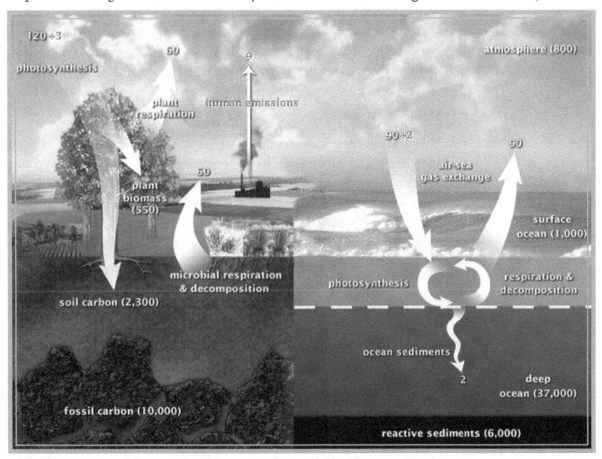

The carbon cycle largely governs global surface temperature by producing heat trapping gases (like carbon dioxide) in the atmosphere, through the greenhouse effect (Figure 3.6). The carbon on the Earth's surface is delivered to the atmosphere via respiration, decomposition, combustion of fossil fuels, and exchange with carbon dissolved in the oceans. The carbon is removed from the atmosphere by photosynthesis on land and in the oceans (forming biomass) as well dissolving into the oceans. The carbon is stored for longer time periods (millions of years) in the deep ocean and in fossil fuels (ancient sunlight). Overall, human activities are adding an additional 9–10 gigatons (9–10,000 million tons) of carbon to the atmosphere each year. About half of the additional carbon emitted by humans into the atmosphere is absorbed by biomass and the oceans, with the remainder accumulating in the atmosphere (4–5 gigatons per year). Beyond water and carbon, other important biogeochemical cycles include those involving nitrogen, oxygen, phosphorous, and sulfur. In the past two centuries, human activities have also had a significant impact on each of these cycles.

Earth System science focuses on how Earth's major systems interact to produce environmental conditions favorable for life. Legendary British scientist, James Lovelock, described this interaction as a feed-back between life (biology) and its physical and chemical components, such that physical and chemical conditions were maintained to sustain life on Earth. He hypothesized that the Earth is itself a sort of super-organism that maintains an environment favorable for life by balancing complex biogeochemical processes and cycles. The super-organism was named Gaia after the Greek goddess of the Earth. Dr. Lovelock defined Gaia as, "a complex entity involving the Earth's biosphere, atmosphere, oceans, and soil; the totality constituting a feedback system which seeks an optimal physical and chemical environment for life on this planet."

It is important to appreciate the complex interactions within the Earth System, and how these sustain life as we know it. As our understanding of past global climate changes improves through increasingly available data, global climate models are being developed to describe how various forcing factors affect internal system interactions, and compare these with past climate records. The better refined the global climate models in describing past changes, the better we will be in predicting changes in the future. Next, we explore climates of the ancient past and until today.

Past and Present Climates

Forcing factors on Earth's climate change over time, and Earth's major systems (lithosphere, atmosphere, biosphere, and hydrosphere) interact in complex ways, which in turn creates changes within Earth's climate, and therefore in each of the spheres (Figure 3.10). This continual forcing, feedback, and response has occurred throughout Earth's four billion years, resulting in dramatic changes in global climate. Since geological time is so immense, we break down Earth's history into eons, which are divided into eras, which are then further divided into periods, epochs, and ages. Difficult to describe using conventional linear graphics, the magnitude of geological time and major events in the biosphere are often presented using successive linear scaled and spiraling time graphics, such as that presented in **Figure 3.12**. Look closely at these graphics to get a feel for how little time humans have existed—the Holocene Epoch marked the beginning of the Agricultural Revolution. In fact, if we were to compress all time since the beginning of the universe, with the Big Bang on January 1, the Earth would have only formed by about mid-September, the first mammals evolved about December 26, and all of human history occurred during the last half hour of New Year's Eve.

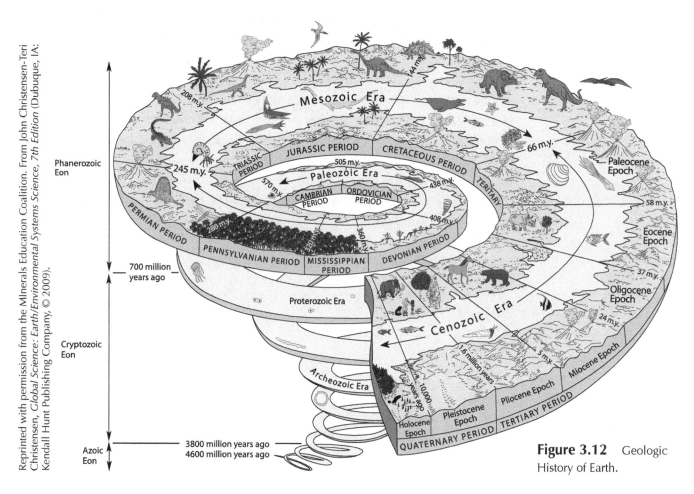

Figure 3.12 Geologic History of Earth.

As evidence of the interaction of Earth's system components, major changes in the biosphere, hydrosphere, and lithosphere during Earth's history have generated significant changes in the composition of the atmosphere and lithosphere. In the early part of Earth, carbon dioxide (CO_2) comprised as much as 30% of Earth's atmosphere by volume. The process known as *chemical weathering* dissolves the carbon dioxide (CO_2) in water (H_2O) making carbonic acid (H_2CO_3), which is used by ocean organisms to grow shells of calcium carbonate ($CaCO_3$), and ultimately forms limestone rock. As plants grew, they too consumed the carbon dioxide in the atmosphere and emitted oxygen (through photosynthesis), which now comprises about 20% of our atmosphere (carbon dioxide is currently less than 1%). The current atmosphere on Earth now has the following approximate composition (by volume): 78% nitrogen (N_2); 20% oxygen (O_2); 1% argon (Ar); 0–3% water vapor (H_2O); 0.04% carbon dioxide (CO_2); and, other trace gases and vapors.

It is beneficial to analyze past climates in terms of average surface temperature, which in large part dictates the operation of the global climate system. This information is often provided to us by *isotope ratios* within cores, drilled out from sediments or glacial ice to preserve the layering in which it was deposited through successive sedimentation or precipitation (snow) cycles (older layers on bottom, younger layers on top). The isotope ratio between oxygen-18 and oxygen-16 ($^{18}O/^{16}O$) within the ice provides us with *proxy* data for the surface temperature in which the ratio was created when it formed into snow. Oxygen-18 (^{18}O) is a heavier isotope than oxygen-16 (^{16}O), the most abundant form of oxygen. Simply stated, water molecules with ^{18}O ($H_2^{18}O$) require more energy to vaporize than that with ^{16}O ($H_2^{16}O$), so when temperatures are warmer, higher $^{18}O/^{16}O$ ratios are measured. A similar process occurs between *deuterium* (^2H) and *protium* (^1H), both isotopes of hydrogen. Isotope ratio data has been collected to describe average surface temperatures on Earth throughout the past 800,000 years or so. **Figure 3.13** shows proxy temperature data from deuterium isotope ratios, as well as CO_2 concentrations from ice cores taken at Lake Vostok in Antarctica. CO_2 concentrations (measured in air bubbles trapped in the ice) are correlated with average surface temperatures throughout this time, indicating the strong connection (correlation) between the Earth's average surface temperature, and its CO_2 concentration.

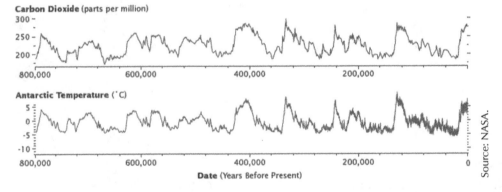

Figure 3.13
Temperature change relative to 1900 (dark lines) and CO_2 concentrations (lighter lines) for past 400,000 years.

Quiz and Problems

(Open Book—Write Answers Below Questions—Show All Work)

Quiz

Answers in Spaces and Show All Work:

1. Name and describe the four (4) basic components of Earth's climate.

2. Historically, what were the three (3) major forcing factors for climate change?

3. Describe the difference between weather and climate.

4. Which is higher above Earth, the stratosphere or the troposphere?

5. Describe how surface temperatures on Earth are maintained.

Problems

Answers in Spaces and Show All Work:

1. In the table below, describe processes in which Earth System Component #1 interacts with Earth System Component #2:

Earth's Component #1	Earth System Component #2	Example Interactions
Lithosphere	Biosphere	
	Atmosphere	
	Hydrosphere	
Biosphere	Lithosphere	
	Atmosphere	
	Hydrosphere	
Atmosphere	Lithosphere	
	Biosphere	
	Hydrosphere	
Hydrosphere	Lithosphere	
	Biosphere	
	Atmosphere	

2. Describe the chemistry of photosynthesis.

3. Sketch and describe a simple ecosystem in terms of its trophic levels.

4

Combustion, Emissions and Environmental Impacts

"To poke a wood fire is more solid enjoyment than almost anything else in the world."

— American Novelist Charles Dudley Warner

combustion and Emissions

Combustion is another term for burning, or fire. Combustion (fire) is a chemical reaction that combines fuel (usually carbon, hydrogen, or hydrocarbons) and oxidant (usually oxygen from air) to make products (exhaust) and thermal energy (heat). Some of the heat produced in the reaction is used to pre-heat the fuel and oxygen, promoting continuous combustion. From campfires to car engines, combustion shares these basic features, as illustrated in **Figure 4.1**. Consider the burning of methane (CH_4), the primary component of natural gas. When the combustion reactants, methane (CH_4) and oxygen (O_2) are combined, the combustion products, carbon dioxide (CO_2) and water (H_2O) are formed. The product molecules are more stable and have lower energies than the mixture of the fuel (methane) and oxidant (oxygen) The difference between the energy of the reactants and products is the amount of heat released during the combustion. However, until there is a spark, or *activation energy*, the combustion does not usually progress.

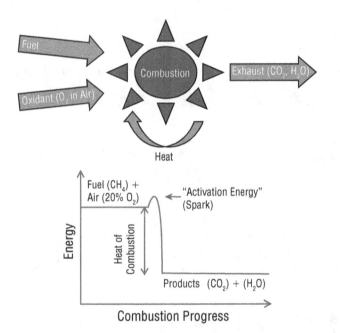

Figure 4.1 Primary Features of the Combustion Process.

Top image © Kendall Hunt Publishing Company; bottom image courtesy of the author.

Combustion processes range from small to large, e.g., lighters to jet engines, with a wide variety of applications including heating, cooling, transportation, and electric power generation. **Table 4.1** compares various common combustion processes in terms of size, rated by typical power (energy/time) levels. Combustion processes (large and small) are ubiquitous, or nearly everywhere, in the modern world. In many ways, combustion (fire) provides a critical underpinning for modern human civilization.

Humans began manipulating fire (the process of burning, or combusting wood or dried dung) over a hundred thousand years ago for warmth and cooking. As humans evolved, we began engineering combustion kilns and furnaces to achieve the higher temperatures required for making ceramics, deriving metals (like iron)

Table 4.1 Common Combustion Processes by Size

Small (W)	Medium (kW)	Large (MW)
Matches / Lighters	Bon Fires / Car Engines	Large Boilers / Jet Engines

from ores, and to produce metal alloys (mixtures of metals) ranging from bronze to steel. In the late 1700s, humans discovered how to use combustion to produce mechanical work (coal-powered steam engine). The steam engine was much more powerful than anything possible before using energy from animals, wind, or water. This discovery ultimately enabled a myriad of useful devices ranging from planes, trains, and automobiles to refrigerators and boilers, or steam generators, which often drive electric generators.

Emissions (exhaust) from combustion include the reaction product gases, along with other oxides, unburned fuel, and ash or any other residue (depending upon the fuel). Common fuels for combustion around the world include dried wood or animal dung, leaves, combustible wastes, and fossil fuels (coal, oil and gas). To better understand combustion reactions we often use simplified descriptions of the fuels, such as: carbon (C)—the bulk of coal or wood (which also contain moisture and many other things); methane (CH_4)—the primary component of natural gas and bio-gas; iso-octane (C_8H_{18})—which roughly describes common gasoline (a primary derivative of crude oil); and, hydrogen (H_2)—from splitting water (separating H_2O using electricity or heat).

The basic combustion reactions for common fuels and their approximate heating values, in units of kilojoules per gram (kJ/g) are presented in **Table 4.2**. The products of the reaction are simply oxidized fuel compounds. In the case of hydrogen (H_2), the product is hydrogen-oxide, or water (H_2O). In the case of pure carbon (C), the product is only carbon-oxide, carbon dioxide (CO_2) when fully oxidized, and carbon monoxide (CO) when partially oxidized. Most fuels are *hydrocarbons*—containing both hydrogen (H) and carbon (C), and therefore producing both water and carbon dioxide when fully burned, or combusted.

During combustion, reactant molecules are transformed into product molecules. For example, a single molecule of methane (CH_4) and two molecules of oxygen (O_2) are transformed via combustion into one molecule of carbon dioxide (CO_2)

Table 4.2 Combustion Reactions for Common Fuels and Approximate Heat Values

Fuel	+ Oxidant	= Products	Heat of Reaction (kJ/g)
H_2	$0.5O_2$	H_2O	142
C	O_2	CO_2	25
CH_4	$2O_2$	$CO_2 + 2H_2O$	55
C_8H_{18}	$12.5O_2$	$8CO_2 + 9H_2O$	47

and two molecules of water (H_2O)—as described in Table 4.2. Although molecules are transformed, the amount (mass) of each element within the molecules remains constant. For example, in the combustion of methane, there are 1 carbon, 4 hydrogen and 4 oxygen atoms in both the reactants (fuel + oxygen) and products. This mass balance of reactions follows from the conservation of mass, and its study called stoichiometry. Stoichiometry is indicated by numbers in front of the reacting or product molecule. For example, burning methane (CH_4) requires 2 oxygen (O_2) molecules and produces 1 carbon dioxide (CO_2) and 2 water (H_2O) molecules. In common terms, stoichiometry for combustion reactions helps describe whether an engine, for example, runs "lean" as compared to running "rich". That is, if there is an over-abundance of air (and therefore O_2) in the combustion (beyond what the reaction in Table 4.2 specifies), oxygen will appear in the exhaust and the combustion is said to have *excess air*, or the engine is running *lean*. If the engine is lacking air, even to the extent where some of the fuel goes unburned and appears in the exhaust, the combustion reaction is said to have *excess fuel*, or the engine is running *rich* on fuel.

In most combustion reactions for hydrocarbon fuels, a small amount of excess air provides the highest efficiency by promoting complete combustion of the fuel. Incomplete combustion results in either fuel and carbon monoxide (CO), or both remaining in the exhaust. The graph in the upper portion of **Figure 4.2** illustrates the mass balance of the fuel + oxygen in terms of exhaust composition and efficiency, and shows typical flames from a Bunsen burner operating under those conditions. As excess air is added to the fuel + air mixture, the fuel usually completely oxidizes, and thereby reduces the fuel and carbon monoxide (CO) in the exhaust. Excess air also results in un-reacted oxygen (O_2) in the exhaust, which is often measured using a sensor, which is used to control the fuel injection system in automotive engines to maximize efficiency, as shown in the lower portion of Figure 4.2.

Most fuels contain much more than simply hydrogen, carbon, or hydrocarbons. They often also contain trace amounts of many other materials, including metals, like lead (Pb) and mercury (Hg), sulfur compounds, like (H_2S), and various non-combustible contents, which form soot, particles, and ash. Additionally, the air is mostly (78%) nitrogen (N_2), and only about 20% oxygen (O_2); therefore, nitrogen is always involved in combustion reactions that utilize air for the oxidant. This means that the product exhaust of many combustion reactions contains several other oxidized compounds, including nitrogen oxide (N_2O), sulfur dioxide (SO_2), and various particulate matter from non-combustible contents. Prompted by the 1990 clean air act, scientific investigations of common combustion products and their impacts on human health lead the United States Environmental Protection Agency (US-EPA) to declare six specific combustion products as criteria air *pollutants*. These six air pollutants, their origins, and health effects are described in **Table 4.3**. More information is at: www.epa.gov/air/urbanair.

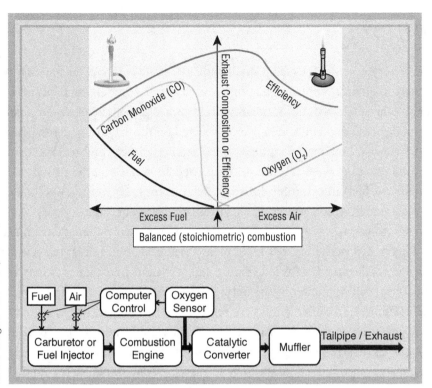

Figure 4.2 Combustion exhaust composition and efficiency as a function of fuel / air ratios (Top) and Automotive engine control system (bottom)

Table 4.3. Six Criteria Air Pollutants per the US-EPA

Pollutant	Origin	Human Health Effects
Carbon Monoxide (CO)	Incomplete combustion of hydrocarbon fuels	Reduces oxygen delivery—can cause death
Nitrogen Oxides (NO_x)	High-temperature combustion with air as oxidant (excess air)	Contributes to formation of ground-level ozone and numerous respiratory effects
Sulfur Dioxide (SO_2)	Combustion of carbon and hydrocarbon fuels containing sulfur (most)	Adverse respiratory effects and reacts with water to form acid rain
Lead (Pb)	Combustion of hydrocarbon fuels with lead and other non-combustion sources	Accumulates in tissues and has adverse neurological effects
Particulate Matter (PM)	Solid particles from fuel and combustion products	Adverse respiratory effects and forms visible haze
Ozone (O_3)—Ground Level	Product of nitrogen oxides reacting with unburned fuels and sunlight	Numerous adverse respiratory effects and smog

Air Pollution Control

In the United States, emissions from small and large-scale combustion systems are often subject to environmental regulations, which aim to limit amounts of the six criteria and other pollutants accumulating in the atmosphere. Therefore, *air pollution control* devices are often installed and operated on the exhaust systems. These include convertors, precipitators, filters, and scrubbers. In car and truck exhaust systems, *catalytic convertors* (Figure 4.2) are used to convert carbon monoxide (from unburned fuel) into carbon dioxide, which is not (yet) a criteria pollutant. In large-scale combustion systems, like coal-burning power plants, a series of air pollution control devices remove particulate matter and other pollutants from the exhaust. **Figure 4.3** presents cross sections of three basic types of large-scale air pollution control devices. On the left (Figure 4.3), a cyclone precipitator system removes heavy particulate matter by separating solids from the gas in a spiral flow. In the middle (Figure 4.3), a filter bag system removes fine particulate matter from the gas, which passes through the filter material. On the right (Figure 4.3), a "packed-bed scrubber" system removes gas-phase contaminants, like sulfur dioxide (SO_2), by forcing the gas through a liquid, which absorbs the contaminants. The liquid is then treated to remove or dilute contaminants, which are then disposed (stored) to avoid atmospheric accumulation.

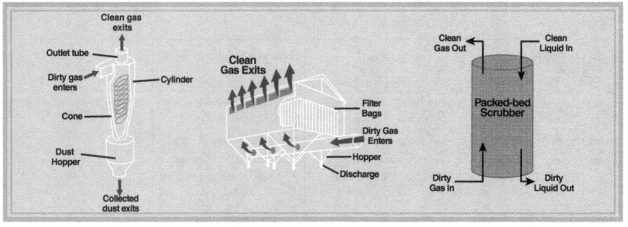

Source: Adapted from EPA: http://www.epa.gov/air/oaqps/eog/course422/ce.html

Greenhouse Gas Emissions

The proliferation of fossil fuel combustion since the late 1700s resulted in steadily increasing CO_2 emissions around the planet. As CO_2 is always a product molecule in combustion of carbon-containing fuels, its emissions have remained largely uncontrolled (relative to the criteria pollutants). Of all of the CO_2 that was emitted into the atmosphere over the past few centuries, less than 50% was extracted from the atmosphere, dissolving into the oceans or converted into biomass through the carbon cycle. The remainder accumulated. That is, in the past few centuries, the rate of CO_2 emissions by fossil fuel combustion has far out-paced the natural carbon cycle to extract it from the atmosphere. The atmospheric concentration of CO_2 is currently higher than it has been in more than eight hundred thousand (800,000) years, and is rising at unprecedented rates.

Figure 4.4 illustrates the CO_2 concentration in Earth's atmosphere over the past 800,000+ years. For the last 10,000 years before the advent of the first practical steam engine, the "Preindustrial" CO_2 concentration in the atmosphere was roughly 278 parts per million (ppm — by volume), and fluctuated very little during this time. During ancient Ice Ages, CO_2 concentrations were as low as about 185 ppm. Thus as a point of reference, the maximum pre-industrial change of CO_2 concentrations was about 100 ppm, and these changes typically required tens of thousands of years. Since the Industrial Revolution, the CO_2 concentration has skyrocketed to over 390 ppm and continues to climb at about 2 ppm per year. The difference between the present and pre-industrial CO_2 concentrations is roughly 40% (>100 ppm), and was established in just over two centuries. This is similar to the differences in CO_2 concentrations during the ice ages compared to those during the relatively stable climate over the past 10,000 years, which enabled the agricultural revolution. That is, CO_2 concentrations have changed about as much in the past 200 years as they have historically in over 20,000 years.

Figure 4.4 Atmospheric CO_2 concentrations throughout 800,000 years.

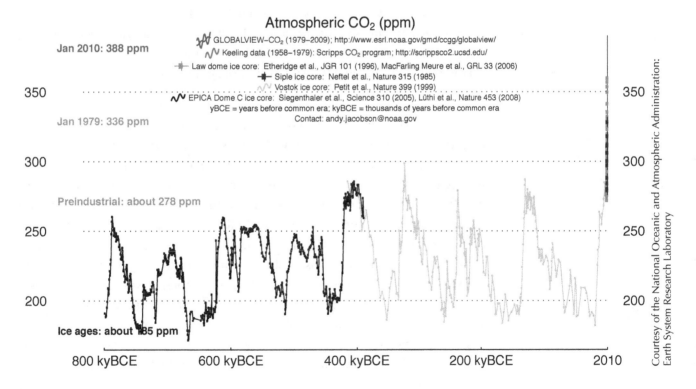

Human-Forced Climate Destabilization

Earth's climate has undergone continuous changes throughout its four billion year history. All of these changes result from the complex interaction among the Earth System components (lithosphere, atmosphere, hydrosphere, and biosphere) as a response (fast or slow) to the natural climate forcing factors (sun's intensity, earth's orbit, and tectonic activity). Since the advent of coal-powered steam engines in the late 1700s, greenhouse gas (namely CO_2) emissions from human activities (compounded by a rapidly increasing global population) have changed the atmospheric chemistry in profound ways.

The rapid rate of atmospheric CO_2 increase over the past two centuries is indisputable. Climate scientists now overwhelmingly agree that the climate forcing factor of atmospheric CO_2 and other greenhouse gases (GHGs) is overriding the pre-industrial forcing factors, and significantly affecting Earth's climate regulation processes. If previous (pre-industrial) climate changes are considered a stable balance of the Earth System, then what industrialized humans are doing is effectively destabilizing the natural climate regulation processes. This is due to the strong interconnection of atmospheric GHG concentration and Earth's average surface temperature (Figure 3.13).

In 1988, the United Nations formed the Intergovernmental Panel on Climate Change (IPCC) to assess the scientific basis for anthropogenic (human-caused) global warming and its impacts on climate. Since then, the IPCC has issued four reports, each articulating the increasing threat of anthropogenic global warming on the Earth System, and human society. Much of the data presented here also appear in publically available IPCC reports. More information is here: www.ipcc.ch.

One way to analyze the forcing factors on Earth's climate is to compare their effective radiative forcing, or how much the factor contributes to solar energy retention or rejection. **Figure 4.5** presents this comparison for nine (9) different radiative forcing components within Earth's climate system for the past century. Positive radiative forcing indicates the amount of additional energy kept within the climate, and produces a general warming of Earth's surface temperatures. Negative radiative forcing indicates the amount of energy blocked from entering the climate, affecting a general cooling of Earth's surface temperatures. On the top are the GHGs, where carbon dioxide (CO_2) dominates, followed by methane (CH_4), nitrogen oxide (N_2O), and halocarbons (hydrocarbons attached to fluorine, chlorine, bromine, or iodine). Next, while the ozone layer in the stratosphere blocks UV, the ground-level ozone in the troposphere (see Table 4.3) has a positive forcing effect. The albedo, or reflectivity, of Earth's surface and aerosols (e.g., vapors, clouds, and contrails from planes) generally have a negative forcing, but have the least certainty due to a relatively lower level of scientific understanding (LOSU) as to their roles in Earth's climate dynamics. While the natural changes solar irradiance has had a slight positive forcing, overall, the effect of anthropogenic (human-caused) forcing dominates the other factors, and is primarily responsible

RF Terms		RF values (W/m²)	Spatial scale	LOSU
Anthropogenic	Long-lived greenhouse gases	CO₂: 1.66 [1.49 to 1.83]	Global	High
		N₂O: 1.48 [0.43 to 0.53] CH₄: 0.16 [0.14 to 0.18] Halocarbons: 0.34 [0.31 to 0.37]	Global	High
	Ozone — Stratospheric / Tropospheric	-0.05 [-0.15 to 0.05] 0.35 [0.25 to 0.65]	Continental to global	Med
	Stratospheric water vapour from CH₄	0.07 [0.02 to 0.12]	Global	Low
	Surface albedo — Land use / Black carbon on snow	-0.2 [-0.4 to 0.0] 0.1 [0.0 to 0.2]	Local to continental	Med -Low
	Total Aerosol — Direct effect	-0.5 [-0.9 to -0.1]	Continental to global	Med -Low
	Total Aerosol — Cloud albedo effect	-0.7 [-1.8 to -0.3]	Continental to global	Low
	Linear contrails	0.01 [0.003 to 0.03]	Continental	Low
Natural	Solar irradiance	0.12 [0.06 to 0.30]	Global	Low
	Total net anthropogenic	1.6 [0.6 to 2.4]		

Radiative Forcing (W/m²)

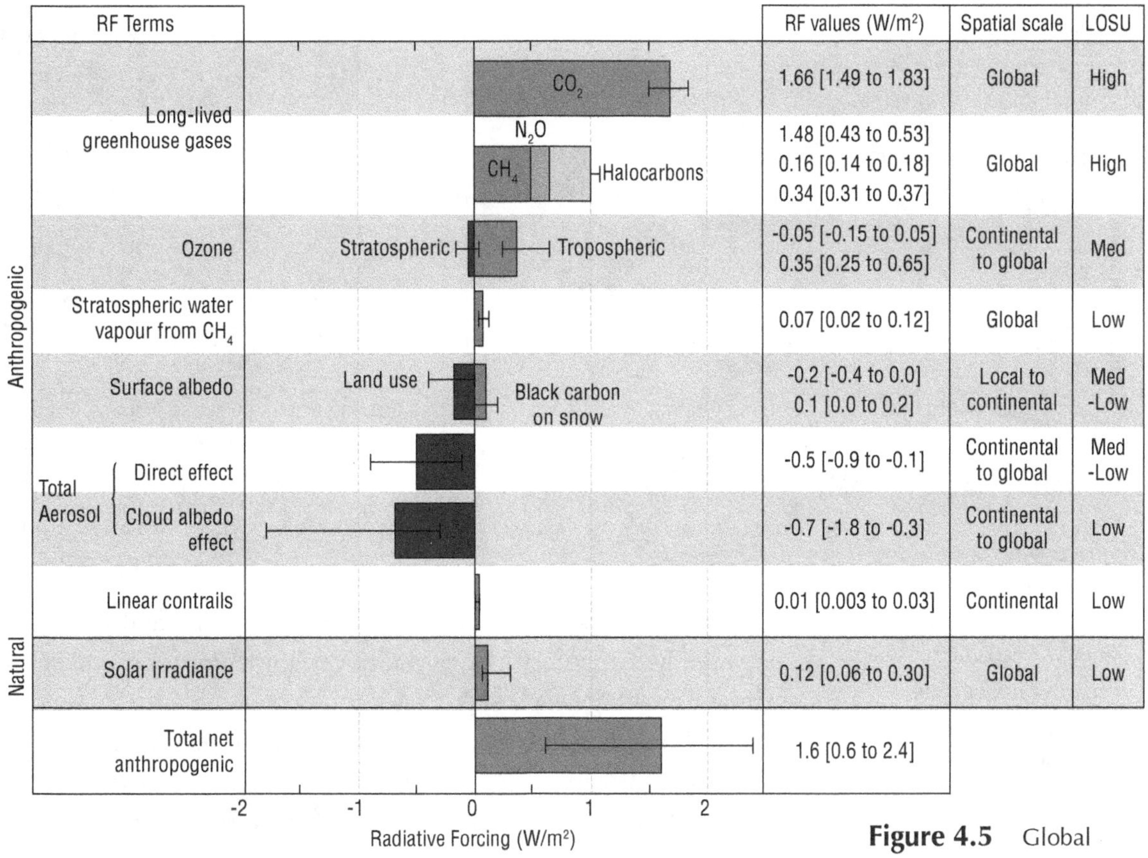

Figure 4.5 Global average radiative forcing (RF) in 2005 by factor, along with spatial scale of impacts and level of scientific understanding (LOSU) of the factors.

for the Earth's rapid warming observed in the past century. The years 2005, 2010, 2011, and 2012 provide additional evidence, as these were among the warmest in recorded history.

The future response of Earth's climate to the increased CO_2 concentration in the atmosphere is difficult to predict. This is especially true as the rate of increase is unprecedented. However, using sophisticated global climate models (GCMs), which accurately describe past climate responses to various known forcing factors, climate scientists can estimate how the climate (Earth's average surface temperatures) will respond to different concentrations of GHG in the future atmosphere. The most basic information in GCMs is the average global surface temperature—Earth's average surface temperatures largely determine ice and sea levels, and the general habitability of different regions on Earth.

Figure 4.6 presents past and predictive average global surface temperatures from years 1900 to 2100. These models are based upon various growth assumptions for CO_2 concentrations in the atmosphere. The year 2000 constant CO_2 concentration is presented for comparison purposes only; even if no additional carbon dioxide-emitting devices ever were built, CO_2 concentrations would continue to grow to at least the low growth emissions scenario (B1). Between the low growth and high growth (A2) assumptions, surface temperatures are predicted to increase 2–6°C by the end of the century. If we realize the upper end, it will have been over a million years ago since the Earth's average surface was that warm.

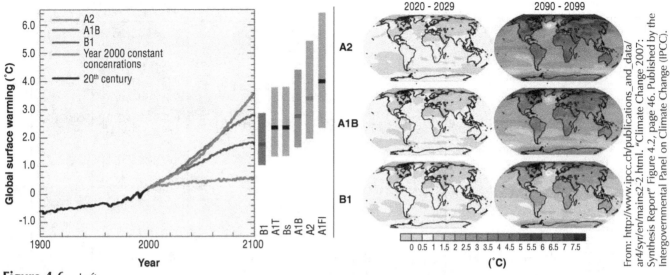

From: http://www.ipcc.ch/publications_and_data/ar4/syr/en/mains2-2.html. "Climate Change 2007: Synthesis Report" Figure 4.2, page 46. Published by the Intergovernmental Panel on Climate Change (IPCC).

Figure 4.6 Left: global surface warming projections based on different emissions scenarios and climate response predictions, along with projection ranges. Right: projected surface temperature changes for 2020-2029 decade, and the 2090-2099 decade relative to the 1990's for the A2 (top), A1B (middle) and B1 (bottom) emissions scenarios and climate response models.

The risks and impacts of global warming are both massive and subtle. Some changes may even be temporarily beneficial to certain regions, e.g., extending growing seasons and opening new trade routes through previously frozen waterways. However, overall the potential impacts are considered to be very serious for life as we know it. Scientists agree that the impacts will depend on the temperature increase and the Earth System response; however, a range of impacts is anticipated with warming in different categories:

- Frequency and severity of extreme weather events.
- Threatened and endangered species face exceedingly challenging existences.
- Ecological, economic, and social systems become susceptible to turmoil as pressures on ecosystem services exceed economic and social demands.
- Irreversible changes if temperatures rise to trigger positive feedbacks, causing uncontrolled warming.

While it is important to understand the severity and urgency of the situation at hand, we must maintain focus on strategies to both adapt to, and mitigate, dramatic climate changes in lieu of ruminating on the possible negative outcomes.

To better understand the challenge, we need to know where the additional GHG in the atmosphere comes from. **Figure 4.7** presents a simplified anthropogenic GHG emissions flow chart for Earth, as prepared by the World Resources Institute; more information at: www.wri.org. On the left side of the chart are the various sectors of GHG emissions, identified on a percentage basis. Dominated by energy through combustion of various fuels, other major emission sectors include land-use change, agriculture, and waste. Moving left to right on the chart, emission sectors are divided into the numerous end use or activities responsible for the emissions, which include three primary anthropogenic GHG gas molecules, carbon dioxide (CO_2), methane (CH_4) and nitrous oxide (N_2O), in addition to various human-made refrigerant fluids (HFCs, PFCs, and SF_6). This breakdown helps identify critical areas for GHG emissions, which are not entirely combustion-related.

World GHG Emissions Flow Chart

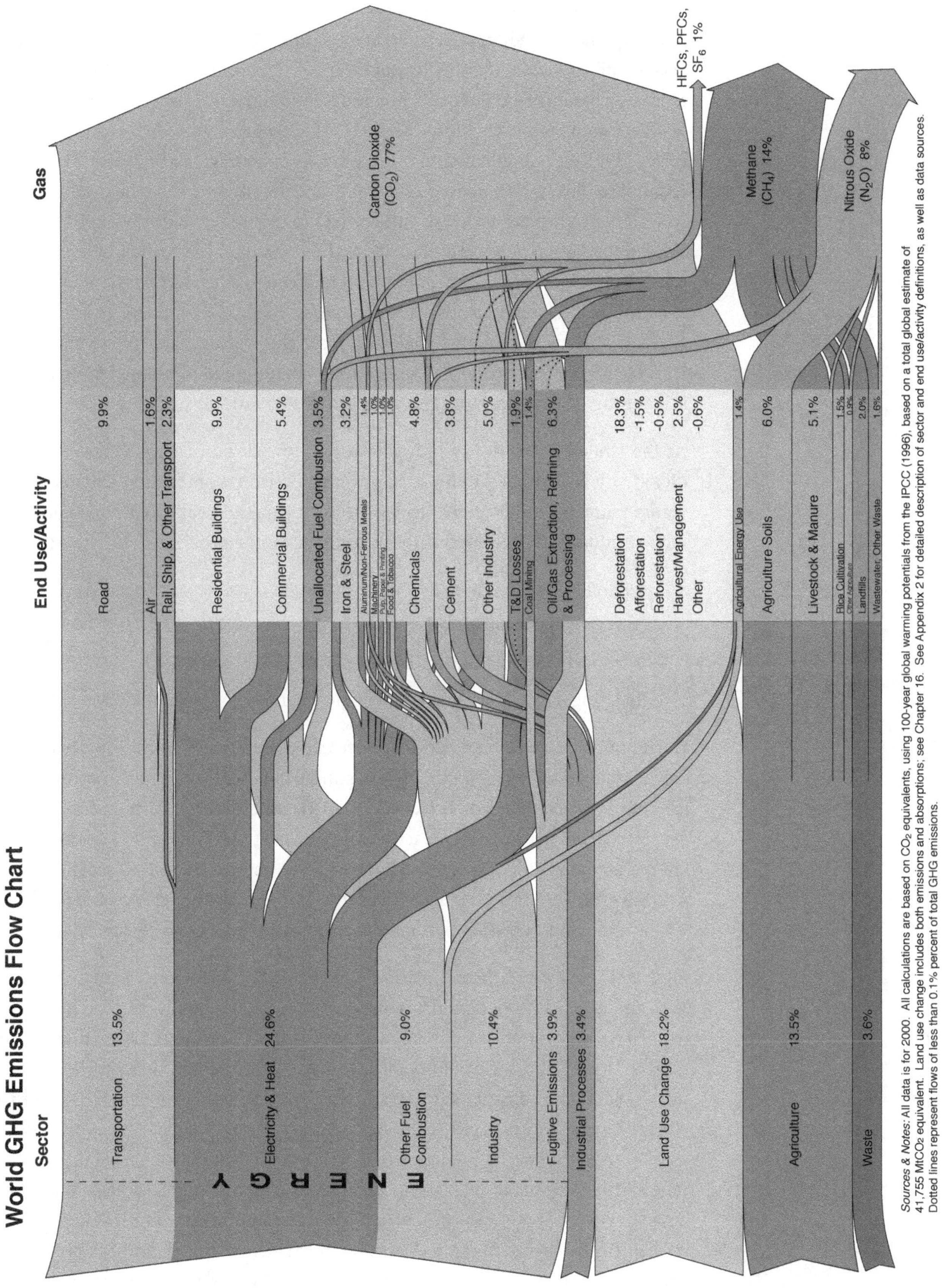

Figure 4.7 World GHG Emissions Flow Chart

Sources & Notes: All data is for 2000. All calculations are based on CO_2 equivalents, using 100-year global warming potentials from the IPCC (1996), based on a total global estimate of 41,755 $MtCO_2$ equivalent. Land use change includes both emissions and absorptions; see Chapter 16. See Appendix 2 for detailed description of sector and end use/activity definitions, as well as data sources. Dotted lines represent flows of less than 0.1% percent of total GHG emissions.

Recently, the United States Department of Justice ruled that carbon dioxide adversely affects human health through its role in anthropogenic global warming, and as such, must be treated as a pollutant. Pollutant emissions are regulated by the US Environmental Protection Agency (EPA) through the 1990 Clean Air Act. Currently, the EPA is implementing programs to encourage carbon dioxide emission reductions through increased efficiency, and tracking carbon dioxide emissions from large industrial sources. These GHG emissions inventories will be used to benchmark progress toward dramatic reductions required for "climate neutrality" removing anthropogenic forcings. Often referred to as *carbon emissions*, GHG gases include CO_2, but also a host of other human-made GHG gasses, including methane and various refrigerants (halocarbons). The tracking and reporting of GHG emissions by individuals and organizations can be complex, but generally-accepted accounting systems are readily available, and increasingly accurate.

As the United States slowly moves toward a low-carbon energy economy, carbon dioxide may even need to be actively removed from combustion exhaust and diverted from the atmosphere. This process is generally called *carbon capture and sequestration* (CCS), and refers to the four sequential processes:

1. separating carbon dioxide (CO_2) from exhaust;
2. compressing it into a supercritical fluid;
3. transporting it to a centralized site; and,
4. injecting it into a designated geologic reservoir for permanent storage.

Current proposals are being developed to explore both geological and oceanic storage of compressed carbon dioxide. Existing technological demonstrations of CCS are common in enhanced oil recovery (EOR), which utilizes carbon dioxide (often separated from natural gas extraction and purification). The carbon dioxide is then compressed and injected into oil reservoirs to change the oil's viscosity and enhance its recovery from the reservoir. Other projects are being explored to develop CCS for the sake of testing permanent carbon storage (not solely for EOR).

Figure 4.8 illustrates how CCS may be integrated with conventional technologies, and also acknowledges the natural carbon sequestration afforded by biomass (terrestrial sequestration). In CCS systems, carbon dioxide (CO_2) is first separated (captured) from the exhausts of processes burning fossil fuels, such as electric power plants and cement factories, and then transported via pipeline to centralized storage. From there, it is either injected into oceanic or geologic reservoirs, used in EOR or other industrial processes (such as a solvent for plastic recycling), or combined with minerals to make solid carbonates, like calcium carbonate, a common antacid. All CCS alternatives have energy input and other technical and political challenges; however, it may present one of the better options in terms of permitting continued fossil fuel use while minimizing climate damage. More research and large-scale testing is needed before CCS can be qualified as a partial solution.

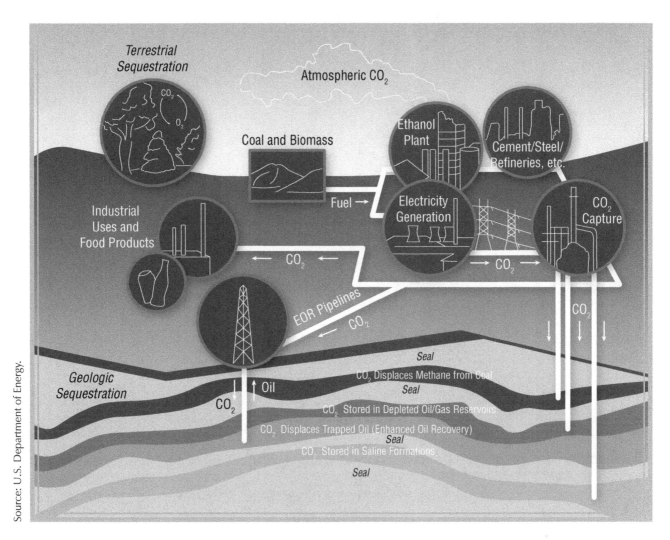

Source: U.S. Department of Energy.

Other large-scale methods of mitigating climate change are often termed "geo-engineering" and include proposals to decrease solar radiation on Earth (e.g., blocking or reflecting sunlight), increase natural carbon sequestration, or both. Many geo-engineering options are attracting increasing interests among scientists and policy-makers around the world.

Figure 4.8 Carbon capture and storage concepts.

Quiz and Problems

(Open Book—Write Answers Below Numbers—Show All Work)

Quiz

Answers in Spaces and Show All Work:

1. What are the primary products of hydrocarbon combustion?

2. What releases more heat; combustion of 10 grams of coal or 10 grams of gasoline?

3. When an engine is running with excess fuel, is it running "lean" or "rich"?

4. What are the six criteria air pollutants?

5. What are the major climate forcing factors and which is the dominant factor today?

Problems

Answers in Spaces and Show All Work:

1. Describe examples of small, medium, and large combustion systems.

2. If 1,000 g (1 kg) of methane (CH_4) is burned every second, how much thermal power is available?

3. In the following table, describe ways to remove or reduce the criteria pollutants from combustion exhaust.

Pollutant	Removal or Reduction Method
Carbon Monoxide (CO)	
Nitrogen Oxides (NO_x)	
Sulfur Dioxide (SO_2)	
Particulate Matter (PM)	

4. If atmospheric CO_2 concentration continues growing at current rates, when will it reach 500 ppm?

5. Define CCS and describe its four major components.

chapter 5

Non-Renewable Energy Sources and Conversion Technologies

"It was the Age of the Hydrocarbon Man….Total world energy consumption more than tripled between 1949 and 1972. Yet that growth paled beside the rise in oil demand, which in the same years increased more than five and a half times over."

— Daniel Yergin—The Prize: The Epic Quest for Oil, Money & Power

Fossil Fuels

At present, fossil fuels comprise more than 80% of total global energy sources, with the remaining coming from nuclear fuels and renewable sources such as biomass, hydro, wind, and direct solar. Fossil fuels are the product of tens to hundreds of millions of years of biogeochemical interactions, and both their physical state (liquid, solid, gas) and chemical content are highly dependent upon when, where, and how they formed. The fossil fuels include coal, oil, natural gas, and methane hydrates, all of which have drastically different histories, abundances/concentrations, and locations around the planet. Essentially, all the fossil fuels store "ancient sunshine," since the original energy input was from the sun over millions of years. Globally, we now use fossil fuels on the rate of several hundred-times their original production rate; that is, it required several hundred years of intense photosynthetic activity to create the fossil fuels which we burn annually. Fossil fuels are primarily used in combustion for heat and mechanical work, with a relatively tiny fraction used in the production of various chemicals, steels, and plastics. Many governments of leading economic nations (including the United States) project that fossil fuels will continue to dominate global energy sources throughout the next several decades. Next, we explore the primary types of fossil fuels, how they formed and how they are used.

Coal

Coal is a black or brownish-black sedimentary rock, comprised mostly of carbon and hydrocarbons, but also containing trace amounts of many other elements and minerals, depending on location and conditions during its formation. Coal was originally plant matter that lived hundreds of millions of years ago, during a period of time when vast swampy forest covered much of Earth. As the plant matter died, it formed peat, which was buried and slowly converted to coal over hundreds of millions of years of exposure to intense heat and pressures within Earth's crust. The coal formation process is presented in **Figure 5.1a**. Coal comes in various types depending upon the conditions under which it formed; some dry, some wet, making for more or less energy content per unit mass. The different types of coal, their relative amounts and end-uses are presented in **Figure 5.1b.**

Coal was the first fossil fuel to be used widely on account of its high energy density, abundance, and ease in extraction. For over a thousand years, coal was used for heating and in the production of metals and alloys like steel, a combination of iron and carbon. By the 1500's its value as a energy-dense heating fuel was well-known, and coal mines were exploiting human and animal power to extract the dense fuel resource. Coal was often burned to boil water into steam, which could be distributed for localized heating. In 1769, the first practical steam engine was invented by James Watt in the United Kingdom (UK), and was employed primarily to aide in coal extraction. Steam engines are powered by steam, produced by coal-fired boilers. The mechanical output (or work) from a steam engine is typically in the form of a rotating shaft or wheel, which then can be used to drive drills, conveyers, pumps, and other equipment used in coal mining and processing. Once the mechanical utility of steam was realized, coal extraction, use, and its associated carbon dioxide emissions, vastly increased—first in the UK, and then around world.

Coal-powered steam engines quickly became the primary driving force behind the Industrial Revolution, which occurred throughout the 1800's and into the early 1900's. This time also witnessed a doubling of world population (from ~1 to 2 billion people from 1800 to 1900). Since then, the population has more than tripled, now at over 7 billion people. Access to abundant and cheap energy in fossil fuels enabled population growth beyond what the Earth's ecosystem services could suport on their own.

Coal is now mined from surface or underground deposits and delivered via trucks, trains, freight ships, or as a slurry in pipes to its consumers, primarily electric power plants and industrial users, such as in iron, steel and cement manufacturing. Coal is commonly measured in tons (2,000 lbs) or metric tonnes (1,000 kg ~2,200 lbs). The global annual consumption of coal is about 8 billion tons, which about doubled from 1980 to 2010, lead by developing countries in Asia. The United States consumes about 1 billion tons of coal annually.

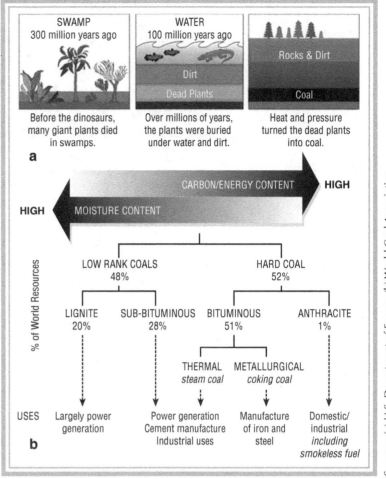

Figure 5.1 Coal formation process (a) and comparisons among types of coal types (b).

Sources: (a) U.S. Department of Energy, (b) World Coal Association.

Oil and Natural Gas

Similar to coal, oil (petroleum) and natural gas formed from organic (living) matter, which thrived tens to hundreds of millions of years ago. Much of the petroleum and natural gas deposits began as diverse ecosystems comprised of various plants and creatures in the vast ancestral oceans. When the living species died, they were successively buried under sand and silt for millions of years. During this time, the dead organic matter was slowly converted by heat and pressure in the Earth's crust, forming energy-dense hydrocarbons, which accumulated in geological reservoirs making several concentrated pockets of various liquids and gases in spotted locations around Earth. In some locations, the hydrocarbons were mixed with dirt and rocks (shale or sand) and deposited over large areas on or deep under Earth's surface. **Figure 5.2** characterizes the formation of petroleum and natural gas.

PETROLEUM & NATURAL GAS FORMATION

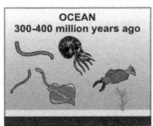

OCEAN 300-400 million years ago

Tiny sea plants and animals died and were buried on the ocean floor. Over time, they were covered by layers of silt and sand.

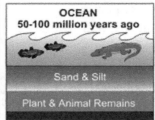

OCEAN 50-100 million years ago

Sand & Silt

Plant & Animal Remains

Over millions of years, the remains were buried deeper and deeper. The enormous heat and pressure turned them into oil and gas.

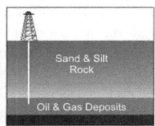

Sand & Silt Rock

Oil & Gas Deposits

Today, we drill down through layers of sand, silt, and rock to reach the rock formations that contain oil and gas deposits.

Figure 5.2 Petroleum and natural gas formation.

To find prospective oil and gas reservoirs, scientists and engineers conduct various surveys of Earth's crust using magnetic, gravitational, and seismic technologies, which help locate and estimate reservoir volumes. This process is known as exploration, and exploratory wells are often drilled to confirm the prospect. Depending upon the type of oil or gas deposit and the technologies used, only a fraction of the reservoir will be recoverable; this is known as oil or gas reserves. Production wells are then engineered to recover the reserves from the reservoirs. Recently-developed horizontal drilling and reservoir-stimulation technologies, including hydraulic fracturing or "fracking", have increased and even enabled oil and gas production from previously unrecoverable reservoirs. A schematic (not to scale) of a hydraulically-fractured natural gas production well is shown in **Figure 5.3**. The steel lined well casing is often bored down thousands of feet through layers in the Earth's crust before being diverted horizontally to enter a hydrocarbon-bearing formation of porous rock (shale or sand). Once into the layer, the well casing can be

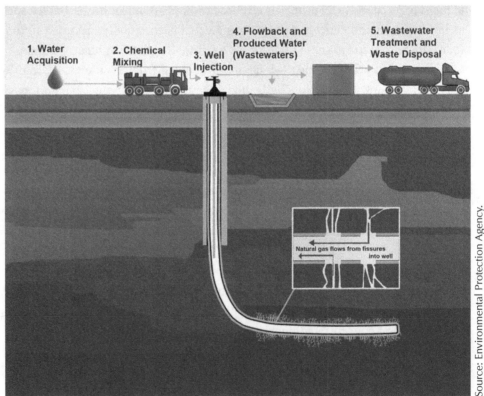

Figure 5.3 Schematic of water cycle in hydraulic fracturing

sequentially perforated from it's "toe", allowing small openings into the well. Next, millions of gallons of water, along with sand and chemicals are pumped from the surface down into the well at high pressure to produce local fractures in the rock formation, allowing gas and oil to enter the well and be driven to the surface. At the surface (well head), the oil/gas are separated from the water, which is stored, treated and disposed, or reinjected into another deep geological layer. Several layers of steel casing and cement are used to isolate the production well contents from that of the layers through which the well bore penetrates.

When petroleum is extracted from a reservoir, it is in a raw form, called *crude oil*. Crude oil compositions and ease in extraction vary from place to place around the world. In some places, oil reservoirs are found only a few hundred feet deep underneath barren deserts and contain high-quality, low-sulphur *sweet* crude. In other places, the reservoirs are thousands of feet beneath the sea floor, which can itself be over a mile deep. Some oil is extracted from non-conventional and low-quality reservoirs, such as oil-laden sand (bitumen) or heavy oil deposits. It is currently estimated that about a third of the oil reserves are in conventional reservoirs, with the other two thirds split between bitumen and heavy oil deposits. Wherever it is extracted and whatever its quality, crude oil is transported to industrial refineries to process and refine the raw material into the various products that we use regularly.

Figure 5.4 presents the processing and refining of crude oil into useful products. Beginning at the left, crude oil from storage is transferred to a furnace where it is

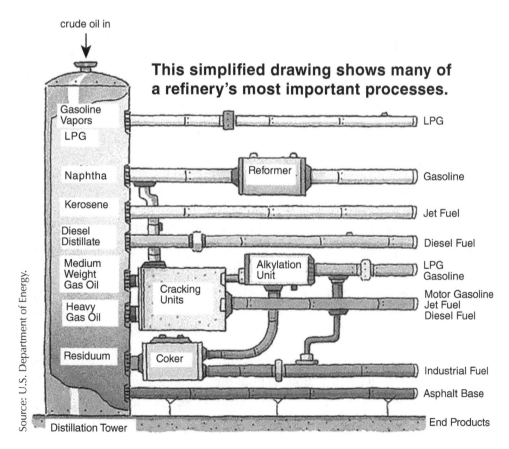

Figure 5.4 Processing and refining crude oil into useful products.

heated and enters a *distillation tower*, which separates the oil into its various fractions of lighter and heavier hydrocarbons. Lighter hydrocarbons rise to the top of the tower and produce fuels such as gasoline, jet fuels, and diesel. Heavier hydrocarbons remain toward the bottom of the tower and comprise products such as asphalt, waxes, and lubricants. This is generally determined by the number of carbon atoms in the molecule comprising the product. The light gases typically have between 1 and 4 carbon atoms, such as 1 in methane (CH_4) or 4 in butane (C_4H_{10}). The lighter fuel liquids have between 5 and 15 carbons, such as 8 in octane (C_8H_{18})—a primary molecule in gasoline or 12 in dodecane ($C_{12}H_{26}$), a primary molecule in diesel fuel. Heavier liquids, including lubricants and greases have between 16-19 carbons, and have high boiling temperatures, making them suitable for gear and crank oils. Hydrocarbon products from crude oil processing that have over 20 carbon atoms are typically solids, like asphalt used in road surfaces.

Hydrocarbons are chain-like, with carbon as the back bone, with hydrogen and other atoms hanging off each carbon. Common hydrocarbons prefixes are described in **Table 5.1**. Using these and various functional groups of atoms or molecules that attach to the basic backbone, we can describe various alcohols, aldehydes, esters, ethers, and amines, from simple one-carbon methane (CH_4) to a straight-chained, eight-carbon octane (C_8H_{18}). Ethanol (C_2H_5OH), the main ingredient in beer, wine, and stronger liquors, is a two-carbon molecule with a hydroxyl (OH) group hanging off one of the carbons—making it an *alcohol*.

Table 5.1 Prefixes for hydrocarbon molecules

Prefix	Meth	Eth	Prop	But	Pent	Hex	Hept	Oct	Non	Dec
Number of Carbon Atoms	1	2	3	4	5	6	7	8	9	10

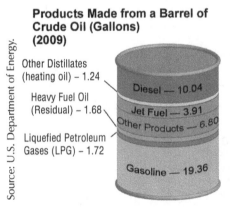

Products Made from a Barrel of Crude Oil (Gallons) (2009)

Other Distillates (heating oil) – 1.24

Heavy Fuel Oil (Residual) – 1.68

Liquefied Petroleum Gases (LPG) – 1.72

Diesel — 10.04

Jet Fuel — 3.91

Other Products — 6.80

Gasoline — 19.36

Crude oil is most often measured in 42 gallon *barrels*, or bbl, and its composition is categorized according to benchmarks to facilitate it global trade. The global daily consumption of oil is about 90 million barrels (80 Mbbl), with the US consuming about 20 million barrels (20 Mbbl) per day. Out of each 42 gallon barrel of crude oil that is refined, the majority is processed into conventional fuels; gasoline, diesel, jet fuel, and heating fuel oil, with the remainder used for other products, including industrial chemicals, asphalt, and plastics. An approximate breakdown for average crude oil is presented in **Figure 5.5**.

Figure 5.5 Breakdown of a 42 gallon barrel of crude oil into useful products.

Figure 5.6 presents a schematic for the natural gas industry, including production, transmission and distribution. In both the oil and gas industry, exploration and production are often considered the upstream processes, whereas the processing, transmission, distribution and end-use are considered downstream processes.

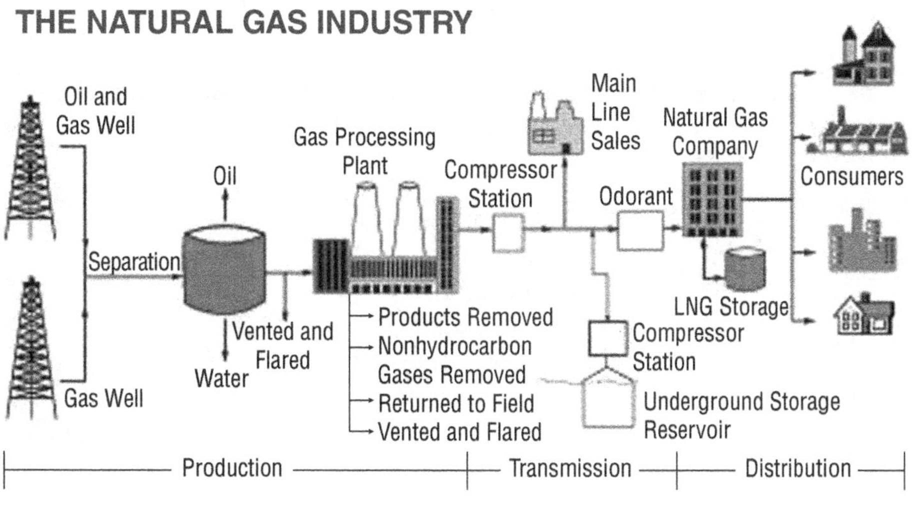

THE NATURAL GAS INDUSTRY

Source: U.S. Department of Energy.

Figure 5.6 Oil and natural gas processing and end uses.

Natural gas is often stored and transported as a compressed gas or a compressed/cooled liquid (LNG) for increased energy density. In the United States, natural gas is typically measured in cubic feet, a specific volume the gas would occupy at standard temperature and pressure. Although the natural gas composition varies, an average energy density is ~1,000 BTU per cubic foot (cf). Larger amounts are measured in therms (100 cf or ~100,000 BTU), Mcf (1,000 cf or ~1,000,000 BTU) and trillion cubic feet, or Tcf (1×10^{12} cf or ~1×10^{15} BTU). The global annual natural gas consumption is about 115 Tcf, with the United States consuming about 30 Tcf. In the United States, natural gas is used about equally three ways: residential and commercial heating; electricity production; and, industrial/chemical use, with a relatively small amount used in transportation. The primary constituent in natural gas is methane (CH_4), which is odorless. So that we can identify leaks, we most often add small but potent amounts of sulfur compounds like H_2S, which smells like rotten eggs.

When coal (or any fossil fuel) is burned, carbon dioxide (CO_2) is always produced. Carbon dioxide (CO_2) emissions are therefore a good measure of the impacts of fossil fuel combustion on Earth's atmosphere. Fossil fuel types (coal, oil or gas) and compositions vary from place to place, but CO_2 is always emitted during its use. Estimated amounts of CO_2 emitted per mass of fuel burned (for electric energy production) are about 1.0 kgCO_2 / kW-hr for coal, about 0.9 kgCO_2 / kW-hr oil and about 0.6 kgCO_2 / kW-hr for natural gas. With a recent influx of natural gas on the market via hydraulic fracturing, its share of the electricity generation market has witnessed substantial increases, often replacing coal and thereby somewhat reducing overall CO_2 emissions from electric power production in the US and elsewhere.

Figure 5.7 presents the total carbon emissions from coal, oil, and gas (in addition to cement manufacturing and gas flares) from 1800 to 2007 in billions of metric tons, or gigatons, of carbon (mostly in the form of CO_2). Historically, coal has been primarily responsible for global carbon emissions due to the spread of coal-powered steam engines. Following World War II, the use of petroleum began significantly contributing to global carbon emissions. Recently, emissions from natural

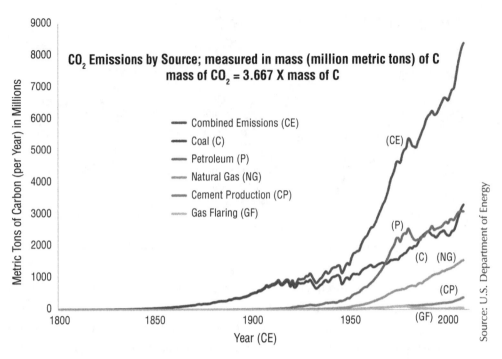

Figure 5.7 Carbon emissions from fossil fuel use 1800-2007.

Source: U.S. Department of Energy

gas combustion began contributing significantly to total emissions, although it remains about half that of coal and petroleum. Cement manufacturing, primarily the high-temperature conversion of calcium carbonate ($CaCO_3$) to cement-forming calcium oxide (CaO) releases carbon dioxide (CO_2) during the reaction (reverse chemical weathering), which is often heated by burning fossil fuels. Gas flares are used to discard *sour gas*, or natural gas containing excessive sulfur compounds, by combustion reactions into less harmful oxides. Gas flares contribute a relatively small, but non-negligible amount of carbon emissions. Total global emissions from fossil fuel combustion are now over 8 gigatons of C, or about 30 gigatons of CO_2, annually.

In terms of total energy used per combustion fuel source, the scene has changed dramatically in the last few centuries. Biomass, e.g., wood and manure, comprised well over 50% of global energy supply until about 1900, when coal became globally dominant. Oil and gas then rose to the top after about 1950, and have remained the predominate source of global energy, comprising ~60% of total energy, along with ~25% from coal and ~10% from both biomass and non-fossil (nuclear and other renewable fuels) combined.

Figure 5.8 presents the approximate breakdown of global energy sources at present, which expands small percentages to identify specific contribution of smaller sources and put them in context of other sources (globally). Oil provides the most energy, followed by coal and gas, and trailed by nuclear and renewable sources; biomass, hydro, direct solar, wind, geothermal, liquid biofuels, and solar electric.

In the United States, the breakdown is not significantly different compared to the global energy supply. **Figure 5.9** presents the history and projections of energy source by type from 1980 to 2035, developed by the US Department of Energy.

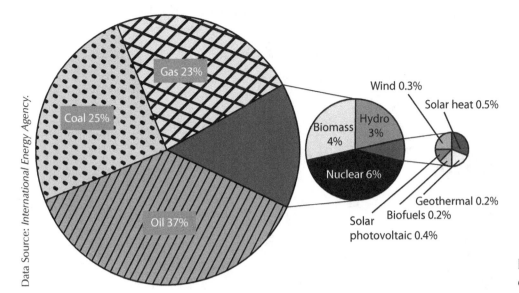

Data Source: *International Energy Agency.*

Figure 5.8 Global energy source breakdown.

Although renewable energy sources are projected to increase substantially in this time, so, too, are our energy demands and use of fossil fuels. Compared with the 2009 baseline, by 2035 the United States is projected to reduce the fraction of fossil fuel energy sources by 5% (from 83% of total in 2009 to 78% of total in 2035). However, since energy demands are also projected to rise from 95 quadrillion BTUs (quads) in 2009 to 115 in 2035, the total amount of fossil fuel consumption is projected to remain about the same, at least according to these projections.

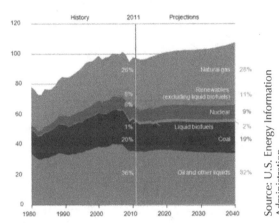

Source: U.S. Energy Information Administration.

Figure 5.9 Primary energy consumption (quadrillion BTU per year).

Methane Hydrates

Methane hydrates (clathrates) are solid compounds, which are comprised of a cage-like ice crystal containing methane, or similar hydrocarbon molecules trapped within. Methane hydrates formed through various microbial processes over millions of years, and are concentrated in trenches along the ocean floor throughout the planet, and within arctic permafrost near the poles. While methane hydrates do not currently contribute significantly to fossil fuel use, the existence of these fascinating structures deserves mention for two reasons. First, the energy resource of methane hydrates is estimated to be on par with all other fossil fuels combined, and could therefore play an increasingly significant role in future energy sources. Second, methane is a powerful greenhouse gas, and its release into the atmosphere, e.g., from melting permafrost, could exacerbate anthropogenic global warming—a positive feedback, or reinforcing mechanism.

Figure 5.10 (left) shows a chunk of methane clathrate burning. As the ice cages melt, methane is released and burned to create heat and continue melting the ice. In Figure 5.10 (right), a cutaway of Earth identifies major methane hydrate resources,

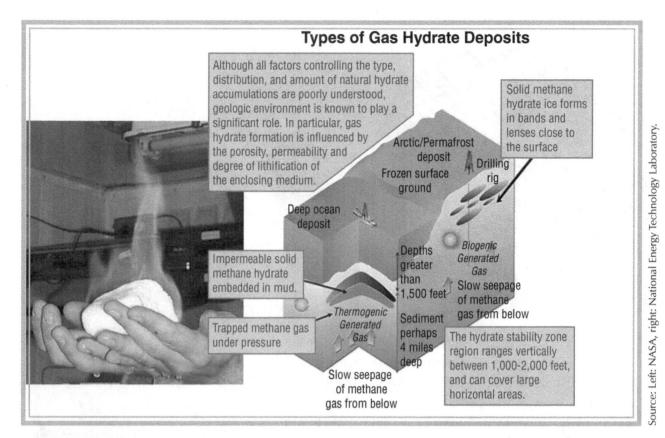

Source: Left: NASA, right: National Energy Technology Laboratory.

Figure 5.10 Methane hydrate combustion (left) and formation characteristics (right).

in both ocean and permafrost locations. New technologies are exploring a substitution of CO_2 molecules for the methane (CH_4) molecules within the ice cage, which may permit a low-carbon source of energy. In this technology, CO_2 is pumped into the methane hydrate formation, where it replaces the CH_4, and the methane is then captured and delivered to the surface. Using water (H_2O) and oxygen (O_2), hydrogen (H_2) can be stripped from the methane (CH_4), with the carbon converted into carbon dioxide (CO_2). In theory, this carbon could then be pumped back down into the methane hydrate reservoir, effectively making them sources of hydrogen and storage sites for carbon dioxide. This, and other methane hydrate exploration and production technologies, will be interesting to follow. Many questions regarding their stability, both in terms of extraction and climate change, remain unanswered and are the subject of intense investigation.

Conversion Technologies

Fossil fuels and other combustibles have served as the mainstay source for practical energy conversion devices. These devices capitalize on combustion to create heat used to power countless mechanical applications, and largely to produce electricity. An overview of some combustion-based devices follows, aiming to familiarize us with these common-place energy conversion processes.

Boilers

Boilers (or steam generators) boil water into steam. Steam has incredible utility, and can be used for transportation, heating, cooling, cooking, cleaning, and more. Boiler systems range from small (household-scale) to enormous (industrial-scale) and vary in their design and operation depending on the application and fuel. Boilers are at the heart of most electric power plants. Regardless of whether the water is boiled by combustion of wood, waste, coal, oil, or gas, or by nuclear reactions, the boiler is designed to efficiently transfer the heat (from the combustion) into the water and create high-temperature steam for its various purposes. **Figure 5.11** presents a generic, mid-sized combustion-based boiler system with unspecified fuel. Water is fed through tubes in a combustion chamber, which boils the water and creates steam. The combustion exhaust (flue) is then used to pre-heat the incoming liquid water.

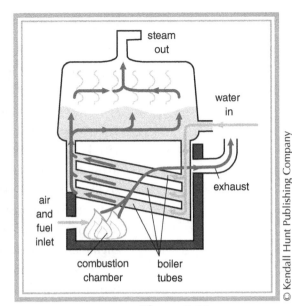

Figure 5.11 Generic combustion-based boiler system.

Furnaces

Furnaces serve a similar purpose to that of boilers, except that they heat air instead of water. Typically, the heating elements within modern furnaces are based on either combustion or electrical resistance, and both make hot air from cold air. Newer, high-efficiency combustion-based furnaces take advantage of converting the water vapor in the combustion exhaust into liquid to extract the *higher-heating* value of fuels. A high-efficiency furnace design is presented in **Figure 5.12**.

In this design, cold air is drawn in at the bottom and passed by metal tubes containing the hot exhaust of the combustion. The combustion exhaust is cooled to the extent it condenses the product water vapor, which is drained. The hot air is then delivered throughout the building via ductwork and vents. The system is typically controlled by feedback from temperature sensors within the rooms in the buildings to adjust how much hot air is allowed in. Furnaces and boilers are primary components within heating, ventilation, and air-conditioning (HVAC) systems (Figure 1.8). Hot air can also be used in drying, cooking and many other useful operations.

Figure 5.12 High efficiency combustion-based hot air furnace.

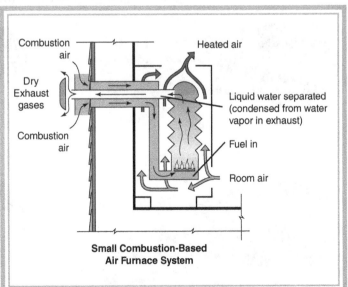

Small Combustion-Based Air Furnace System

Reprinted with permission from the Minerals Education Coalition. Adapted from John Christensen-Teri Christensen, *Global Science: Earth/Environmental Systems Science, 7th Edition* (Dubuque, IA: Kendall Hunt Publishing Company, © 2009).

Engines

Engines capitalize on combustion or other source of heat (often in the form of high-temperature steam) to produce useable mechanical energy, or work. In combustion-based engines, the combustion process typically occurs inside a piston-cylinder arrangement, as shown in **Figure 5.13**. Fuel and air are mixed into the cylinder and combustion is ignited by a spark plug—or simply by compression in diesel engines. As the combustion occurs, it pushes the piston, which is connected to the crankshaft (and other piston-cylinders). In unison, the piston cylinders each go through a four-stroke cycle of: 1) drawing in fuel and air; 2) compressing the mixture; 3) combusting the mixture; and, 4) exhausting the product gases. This is then continued several times a second for each of the engine cylinders. Although engine design varies significantly depending upon application and fuel, these are the common features for all combustion-based engines. When weight is an issue, two-stroke engines combine the first and last two steps into one, and thereby have more power per weight to be used in lawnmowers, chainsaws, and snowmobiles. However, they also are noisy and less efficient than their four-stroke counterparts. Additionally, lubricant must be added to the two-stroke fuel, which produces dirty exhaust. Diesel-fueled engines run similarly, except the compression pressure is sufficient activation to initiate the combustion, and therefore, they do not require sparkplugs.

Figure 5.13 Cross section and operation of a basic four-stroke gasoline engine.

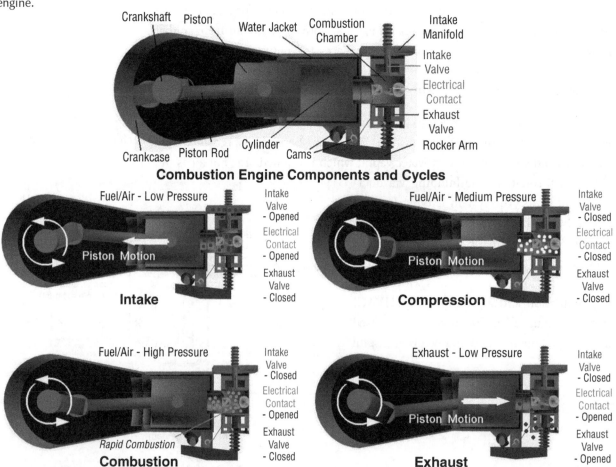

Source: NASA.

Steam-based engines differ in that they utilize steam as the working fluid as opposed the fuel + air mixture. As high-pressure steam enters a piston-cylinder arrangement, it pushes the piston, which is connected to a fly wheel, and exhausts low-pressure steam in the process. As the fly wheel rotates, it pushes the piston back into its original position, and the high-pressure steam enters to repeat the cycle. Modern steam engines operate in similar fashion, but in relatively compact form. Other engines can operate by cycling a working fluid heated by external heat sources, such as concentrated sunlight, or external combustion. In all heat-powered engines, thermal energy (heat) must be transferred from high-temperature source to low temperature source, creating usable work. The biggest the temperature difference between the sources, the higher the energy conversion efficiency (T>M). Unfortunately the high-temperature source is limited by materials degrading (e.g., corrosion/melting), and the lower-temperature source is typically the temperature of air or body of water. Due to these constraints, modern heat engines are typically limited to much less than 50%.

Steam Turbines

Steam turbines are another form of heat engine, driven by high-velocity (high-temperature) steam. Steam turbine engines take advantage of the thermal energy within steam, which is converted to mechanical energy as it moves turbine blades connected to a central shaft, as depicted in **Figure 5.14**. Often, a water- or air-cooled condenser is used to provide a low-temperature source of heat to help collapse the steam from vapor to liquid and drive the process. Modern steam turbine engines have significantly more complexity, often with several stages of various size to convert as much of the steam's thermal energy into usable mechanical energy (in the form of a spinning shaft). Most often the spinning shaft is connected to an electromechanical generator to produce electricity (electrical energy) from the mechanical shaft work.

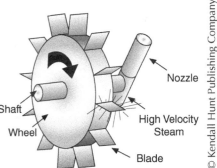

Figure 5.14 Basic steam turbine engine operation.

Gas Turbines

Combustion-based turbines (or jet engines) helped revolutionize transportation in the mid-1900s. They operate on the catchy cycle: "suck; squeeze; bang; blow". First, air is drawn into the turbine and compressed via fans and compressor blades (suck and squeeze). Then, the compressed air is mixed with the fuel and combustion (bang) imparts thermal and mechanical energy onto the turbine blades, which are connected to a central shaft that drives the air compressor and intake fan. The combustion exhaust and bypass air are then forced out of the back of the turbine (blow) causing a forward thrust. A generic turbine engine is presented in **Figure 5.15**. These systems are highly efficient, and now are used for both jet-plane transportation and land-based electric power generation applications.

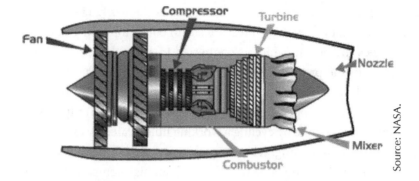

Figure 5.15
Combustion-based turbine engine cross section.

Source: NASA.

Electric Power Plants

Most electric power plants are steam- and/or gas-turbine- powered, with the steam coming from boilers burning fossil or nuclear fuel, or from the hot exhaust from the gas turbine. The fossil-fuel-powered turbine spins a shaft at high speed, which is connected to the electromechanical generator (**Figure 5.16**). So long as the turbine spins its shaft, electric energy is available. As more electricity is needed, more fuel is fed to the boiler so that more steam can enter the turbine and provide the electricity on demand. Often, power plants operate with "spinning reserve" to accommodate large electrical load fluctuations on the electrical utility grid.

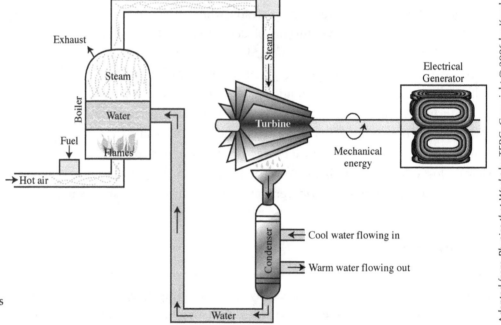

Figure 5.16 Basic steam-powered electric power plant components and operation.

Adapted from *Physics that Works* by TERC. Copyright © 2006 by Kendall Hunt Publishing Company. Reprinted by permission.

Quiz

. . . .

(Open Book—Write Answers Below Numbers—Show All Work)

1. When, approximately, did humans begin using more fossil fuels than biomass?

2. What type of coal produces more energy per unit mass when burned, anthracite or lignite?

3. What is crude oil, where does it come from, and how much do we use daily, globally?

4. What is heavier hydrocarbon, asphalt, or gasoline?

5. What are the four strokes in an internal combustion engine?

6. What are the four steps in a gas turbine engine?

7. Approximately what percentage of crude oil is converted into gasoline in a simple refinery?

8. Describe why natural gas is often liquefied before transport.

9. Describe two reasons why methane clathrates become important.

10. How many carbon atoms does "hexane" have?

6

Atoms and Atomic Energy

"Our children will enjoy in their homes electrical energy too cheap to meter."

— Former US Atomic Energy Commission Chairman Lewis Strauss, 1954

Atoms, Ions, Isotopes and Radiation

Atoms are the basic building blocks of matter. Atoms consist of a core, or *nucleus*, which contains a positively-charged *proton* and a neutral charged *neutron*. The atom's nucleus is orbited by negatively-charged *electrons* equal in quantity to the protons in the nucleus. Breaking it down further we find a multitude of fascinating constituent, "sub-atomic" particles that behave as spinning bundles of energy, concepts beyond the present scope. For a basic understanding, remember the three (3) primary components; *protons* and *neutrons*, making up the *nucleus* and most of the atom's mass, and, the much-lighter *electrons*, swarming around the *nucleus* in shell-like orbits. When electrons orbiting a nucleus shift from one orbiting shell to another, a *photon* with a specific energy (representing the electron shell transition) is either absorbed or emitted by the atom.

An enormous distance exists between the electrons and nucleus in an atom. **Figure 6.1** illustrates this for a helium atom, which has two protons and two neutrons in its nucleus, surrounded by two electrons. The total size of the helium atom, to the extent of its electron shell, is about one Angstrom, or 1×10^{-10} meters. The nucleus is less than one femtometer across, or about 1×10^{-15} meters. This means that the vast majority of matter is space between atomic particles; in fact, if all this space was taken away, all the matter in the entire universe would be about the size of a cherry! If the He nucleus represented a one meter distance (a beach ball), the nearest electron would be ~100 km (about 62 miles) away.

General chemistry mostly concerns the interaction among the outer electrons of atoms, which form bonds between atoms, making simple and complex molecules, and ultimately life (biochemistry). Nuclear chemistry deals with atomic nuclei, their various forms and interactions with other particles, and the tremendous energy stored within them. Nuclear processes include *fission* and *fusion*; fusion is the combination of nuclei making heavier atoms whereas fission is the splitting of heavy nuclei into lighter atoms. Both processes release radiant and thermal energy.

Different, unique atoms are arranged into numbered *elements* depending upon how many protons they have. Hydrogen is the lightest element, with only one proton and one electron—typically no neutrons (atomic number 1). Helium is next, with two protons, two electrons and normally two neutrons (atomic number 2). This continues so on and so forth, with now 118 elements discovered and characterized.

Figure 6.1 Helium atom showing size of the atom in comparison with the nucleus. The two electrons make a cloud-like shell surrounding the tiny nucleus, with two protons and two neutrons.

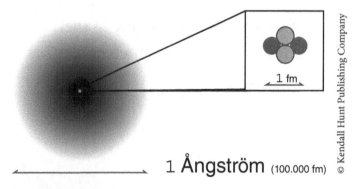

1 fm

1 Ångström (100.000 fm)

© Kendall Hunt Publishing Company

Figure 6.2 presents the standard periodic table of the elements. The symbol within the element box stands for the element's name, e.g., "H" for "Hydrogen", with the atomic number above (proton count), and the number below indicating the atomic mass (in grams, g) per *mol* of atoms (1 mol = 6.02 × 10²³). For example, carbon (C) has the number 12.0 below it, which indicates that 1 mol of C, or 6.02 × 10²³ atoms of carbon, has exactly 12.0 grams of mass, primarily from the 6 protons and 6 neutrons in its nucleus. The elements are first arranged into rows, or periods, according to how many electron shells orbit the nucleus in the atom (up to seven). Hydrogen and helium only have one electron shell, hydrogen with one electron, and helium with two. The next row of elements has two electron shells, and contains the elements lithium, beryllium, boron, carbon, nitrogen, oxygen, fluorine, and neon. The trend continues through to the larger elements, until up to seven orbit shells are filled with electrons.

The elements are further arranged into columns, or groups, according to how many electrons orbit the outer shell in a stable configuration. Because atomic bonding occurs in the outer shell electrons, elements within groups tend to have similar characteristics. Depending upon the configuration of the nucleus and electron shells, the elements are separated into various metals, non-metals, halogens, and inert gases, which have similar chemical properties. The elements, BrINClHOF, pronounced "brinkel-hoff", usually exist as diatomic (2-atom) molecules, such as Br_2, I_2, Cl_2, F_2, H_2, O_2 or N_2. The elements CHNOPS, pronounced "chin-ops" (carbon, hydrogen, nitrogen, oxygen, phosphorous, and sulfur), are the basic elements required for life as we know it. However, most life contains dozens of elements in varying presence.

When an atom or molecule has an excess or deficiency of electrons relative to its standard configuration, it acquires an overall negative or positive charge, and is called an *ion*. The process of stripping or adding electrons from or to an atom is called *ionization*. Ions are used in batteries and many other electrochemical devices, as they can transfer electrical charge. Ionization can also occur when radiation interacts with an atom. Radiation comes in packets, called *photons*, which behave simultaneously as waves and as particles of energy.

The electomagnetic spectrum in **Figure 6.3** shows the wide range of photon wavelengths (λ), usually measured in nanometers (nm, 1 nm = 1 × 10⁻⁹ m), and frequencies (v), usually measured in Hertz (Hz, 1 Hz = 1/second = 1 s⁻¹). When multiplied, the photon wavelength and frequency make the speed of light (c), which is approximately 300 million meters per second, or 300 × 10⁶ m/s (Equation 6.1).

$$\lambda \times v = c$$

Equation 6.1

The energy (E) of the photons can be determined by multiplying the Planck Constant (h) with photon frequency, as in Equation 6.2. The Planck constant (h) is

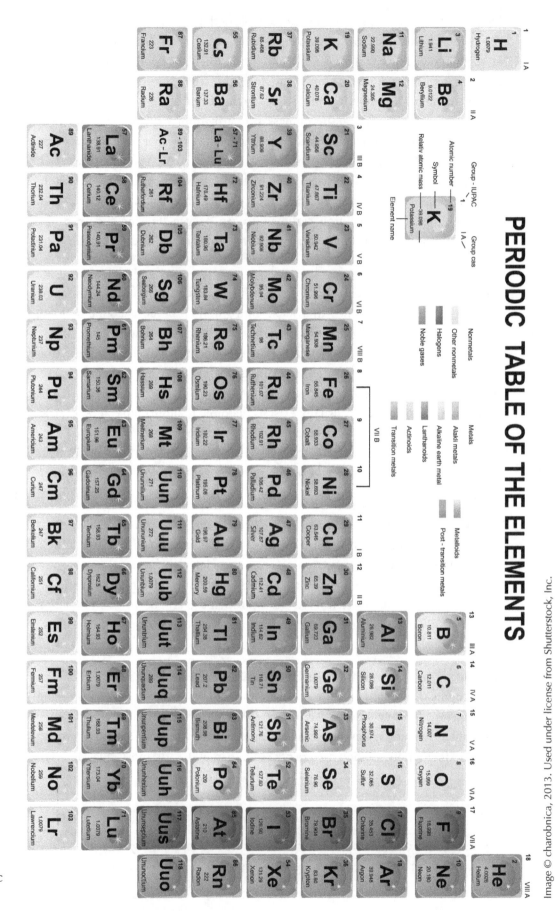

Figure 6.2 Periodic table of the elements.

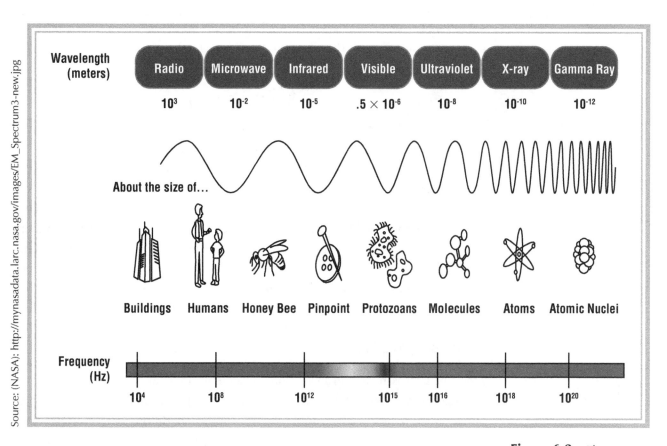

Source: (NASA): http://mynasadata.larc.nasa.gov/images/EM_Spectrum3-new.jpg

approximately equal to 6.626×10^{-34} J-s; a small number, but when multiplied by photons with high frequencies, large energy results.

Figure 6.3 The electromagnetic spectrum with common wavelengths, frequencies, and visible range colors.

$$E = h \times v$$

Equation 6.2

Using Equations 6.1 and 6.2 and some algebra, the Energy (E) can also be determined by the wavelength, as in Equation 6.3.

$$E = (h \times c) / \lambda$$

Equation 6.3

At large wavelengths (low frequencies), like radio and microwaves, the energy is low and the photons have little ionization affects, but are useful in communications and heating. In the mid-lower wavelengths (mid-upper frequencies) photons can induce photo-chemical processes, like photosynthesis, and photo-electrical processes, like solar-electric panels, and also some ionization. At the lowest wavelengths (highest frequencies), the photons have very high energy and ionization potential, and can penetrate deep into most materials. In nuclear processes, such as fusion and fission, tiny amounts of mass (m) are converted into large amounts of radiant (electromagnetic) energy (E) through the famous Equation 6.4. By mutliplying the mass by the speed of light-squared ($c^2 = 9 \times 10^{16}$ m^2/s^2) tremendous amounts of energy are released in nuclear fusion processes (like in our sun) and nuclear fission processes (like in nuclear power plants).

$$E = m \times c^2$$

Equation 6.4.

When an atom has an excess or deficiency of neutrons, relative to its standard number in the periodic table, it is called an *isotope*. Isotopes of elements have neutral charge, but their mass is different compared to the basic element. Atomic mass is predominately from protons and neutrons with much lighter electrons in orbit. In fact, the number of protons and neutrons add to make the *atomic number*. For example, the most stable form of uranium (U), atomic number 92 (the number of protons), has atomic mass of about 238. Therefore, the number of neutrons in uranium-238 is: 238 (mass) − 92 (protons) = 146 (neutrons). A common isotope of uranim used in nuclear power plants is uranium-235, which is lighter than the stable uranium atom and has 235 (mass) − 92 (protons) = 143 (neutrons). This combination makes U-235 unstable and *fissile*, readily breaking into smaller isotopes and releasing energy.

From the approximately 110 well-known elements, there are over 1700 stable and unstable isotopes with different nuclei. Stable isotopes do not change with time, but their ratios are a function of environmental conditions. For example, the isotopic ratios of oxygen-18 (^{18}O) to oxygen-16 (^{16}O) or deuterium (^{2}H or D) to hydrogen (H) in ice (H_2O) is representative of temperature during which the ice was formed. Some isotopes *radioactively decay* into more stable isotopes steadily with time. This decay process is so precise, that it can be used to accurately set clocks and date rocks, ice, and artifacts. For example, the ratio of carbon-14 (^{14}C) or carbon-13 (^{13}C) to carbon-12 (^{12}C) can identify the age of carbon-containing materials through *radio-carbon dating*. As carbon is buried (taken out of the carbon cycle in the atmosphere) the carbon-14 (originally produced in upper atmosphere from nitrogen interaction with cosmic rays) steadily decays into carbon-13 and carbon-12 reaching half its original concentration (*half-life*) in 5,730 (+- 40) years. By measuring the ratios of carbon isotopes, we can accurately determine how old the carbon-containing material is, up to about 60,000 years old (after this, the carbon-14 content is difficult to measure).

Figure 6.4 presents a schematic (not-to-scale) of the carbon atom (left) with the different stable isotopes (right). Another example of carbon-dating is that of

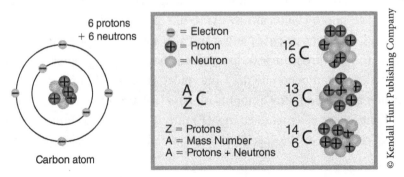

Figure 6.4 The carbon atom (left) with three carbon isotopes (right).

© Kendall Hunt Publishing Company

atmospheric carbon—to determine if it came from burning fossil fuels, which have almost negligible amounts of ^{14}C (due to is decay while being buried for millions of years). Indeed, much of the atmospheric carbon is directly from fossil fuels, providing another line of evidence for impacts of human activities.

The radioactive decay process emits various forms of ionizing particles and/or electromagnetic radiation steadily with time. In addition to using this radioactive decay processes to determine the age of rocks and artifacts, we have found ways to extract and concentrate radioactive isotopes, such as uranium-235 (^{235}U), and use them as sources of intense thermal energy (for boilers and bombs). Radiation can be powerful, and penetrate various materials depending on its form. **Figure 6.5** presents different types of ionizing radiation, and their penetration potentials in different materials. *Alpha radiation* is weak, and can be stopped by a human hand. *Beta radiation* can penetrate a human hand, but is stopped by even thick aluminum. X-Ray radiation, as used in medicine, can penetrate humans and aluminum, but is stopped by lead —the reason why protective lead aprons are typically used in X-Ray procedures. Gamma ray radiation can penetrate through thin lead, and often requires thick lead or even leaded-concrete to stop it. Fast-moving neutrons are also considered radiation, and can penetrate through lead and even thick concrete, depending upon their energy. Neutrons are involved in nuclear reactions, both fusion and fission, and are now used to investigate materials on a sub-atomic level. More information on the latter is at: www.neutron.anl.gov.

Radiation is often measured with the basic unit of a Curie (Ci), which describes the number of disintegrations of a radioactive material per unit time. Depending on the type of radiation, various amounts of energy are associated with each disintegration event (e.g., whether the disintegration produces alpha, beta and/or gamma ray radiation). Radiation absorbed dosage (*rad*) is measured in units of energy per unit mass of absorbing material. However, different tissues are more sensitive than others to radiation, so a *dose equivalent* is used to determine its biological impact. The dose equivalent uses the rad with a multiplication factor depending upon the type of radiation and the tissue its effecting. This is measured in units of *Sievert* (Sv), or more often *millisieverts* (mSv). Most of the radiation dosage humans receive is from

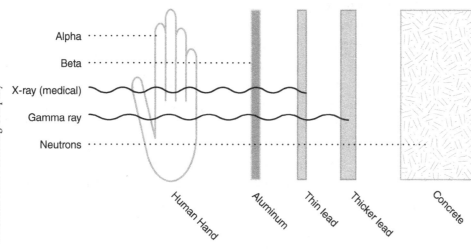

Figure 6.5 Penetration Potential of Different Ionizing Radiation

naturally occurring sources in both space and in the Earth. **Figure 6.6** presents the average effects of ionizing radiation dosage on people. The fatal radiation dosage is about 10,000 mSv, whereas adverse health effects are realized at about 1,000 mSv. Most common exposures in medical, outdoor play, and even working in the nuclear industry are well below 1 mSv. Depending on where you live, the average annual ionizing radiation dosage is about 2–3 mSv.

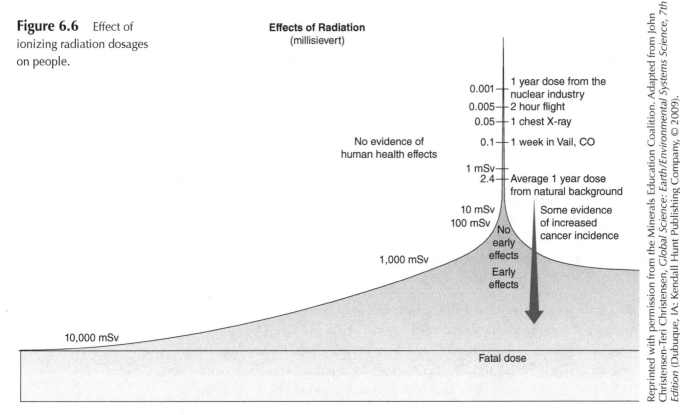

Figure 6.6 Effect of ionizing radiation dosages on people.

Reprinted with permission from the Minerals Education Coalition. Adapted from John Christensen-Teri Christensen, *Global Science: Earth/Environmental Systems Science, 7th Edition* (Dubuque, IA: Kendall Hunt Publishing Company, © 2009).

Nuclear Fission and Nuclear Power Plants

Fission is the splitting of atomic nuclei into lighter nuclei, and the associated release of energy. Fission reactions are constantly underway within Earth's crust, as unstable isotopes of heavy elements, such as uranium, decay into more stable, lighter nuclei, and release neutrons radiant energy in the process. Nuclear fission for electric power works by simply boiling water into steam from the energy released in the decay of concentrated uranium-235 (^{235}U) fuel pellets. Uranium-235 has extremely high energy density, about 80 million kJ per g, which dwarfs that of coal, at about 25 kJ per g. The nuclear fission process in uranium-235 can become a sustained *chain-reaction* if there is a *critical mass* of the fuel available. Commercial uranium-235 fuel pellets contain about 3–5% ^{235}U, with the remainder being mostly stable uranium-238 (^{238}U). On average, about 3 neutrons are produced during fission of 1 ^{235}U nucleus. If other nearby ^{235}U nuclei absorb the neutrons, additional fission reactions ensue and about 3 more neutrons are produced for each. This cascading, chain-reaction (shown in **Figure 6.7**) can be terminated by neutron capture from non-fissile materials, such as stable ^{238}U nuclei. The reaction is controlled by neutron capturing control rods, and moderated in water or graphite to slow down the faster neutrons. As the ^{235}U fuel decays, a multitude of radioactive and stable isotopes accumulate. After some time, the fuel becomes depleted and the spent fuel goes into storage or *reprocessing* to recover unspent fuel and other useful isotopes.

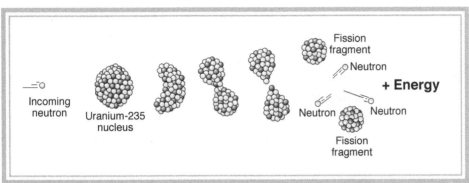

Figure 6.7 Nuclear fission of ^{235}U

Uranium-235 occurs naturally in small concentrations (<1%) in ores found various places around the world. Since uranium ore reserves are finite in nature, nuclear fission is considered non-renewable. However, nuclear fission does not emit greenhouse gases during operation, its waste can be recycled, and there are substantial fission fuel (^{235}U and others) reserves world-wide. Furthermore, fission reactions do not require oxygen, making them suitable for powering submarines and satellites.

The basic features of nuclear fission-based electric power production are presented in **Figure 6.8**. Within a reinforced containment structure, a fission reactor contains the fission fuel (3-5% ^{235}U) within a pressure vessel, along with an external blanket

Figure 6.8 Nuclear fission-based electric power plant basics.

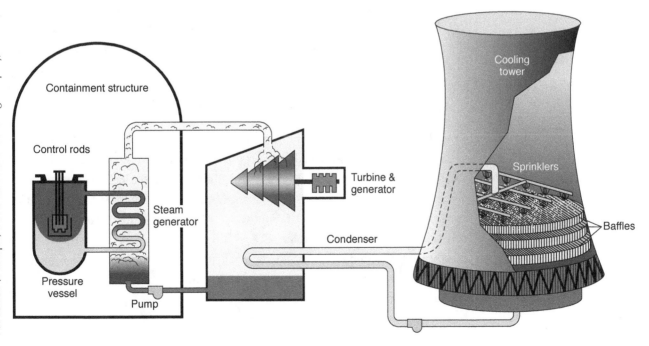

and internal control rods made of neutron-absorbing materials. As the fission is controlled in a critical state, a steady output of thermal energy heats a thermal fluid, which is used to heat water into steam through a *heat exchanger*. The steam then powers a steam turbine connected to an electromechanical generator.

There are three modes of nuclear fission chain reactions; *sub-critical*, *critical*, and *super-critical*. *Sub-critical* nuclear fission ultimately dissipates, i.e., less than one neutron on average produced in the initial fission is able to cause fission in other nuclei. This is often the case when fission fuel is not sufficiently concentrated with fissile nuclei. *Critical* nuclear fission sustains itself, i.e. just enough neutrons produced in fission reactions go on to cause additional fission, as in the case of commercially-operating fission reactors. In *super-critical* nuclear fission, the chain reaction continuously accelerates until all of the fuel is depleted. If the fuel is commercial-grade (3-5% ^{235}U) and the nuclear fission chain reaction becomes uncontrolled and super-critical for extended periods, it can cause the fuel to melt through its containment, causing a "melt-down", such as in the Chernobyl incident in 1986, or the more recent event in Fukushima in 2011. These events released the radioactive fuel and waste products into the environment, which will persist for thousands of years before decaying. The areas nearby these yet-unfolding disasters may remain uninhabitable for centuries, and the regional and global effects are largely unknown. The Chernobyl incident is blamed for over 4,000 deaths associated with the initial explosion/fires and subsequent radiation exposure. The Fukushima incident has not yet been directly blamed for human casualties on account of radiation exposures.

To prevent these disasters, nuclear reactors are heavily regulated and equipped with modern computer-controlled and fail-safe operation, with redundant back-up systems. Additionally, multiple layers of structural security are built into modern nuclear fission reactors to contain the process and radioactive materials. An illustration of these security layers is shown in **Figure 6.9**. These layers protect against both external forces and radiation escaping into the environment. Regardless of the protection, some risk always remains with any energy conversion technology and in particular with nuclear fission, which produces highly-radioactive and environmentally-persistent wastes. Some liken the risk to sky-diving; it is fairly safe if performed correctly, but accidents can have extreme consequences.

In spite of the inherent risks, nuclear fission technology is relatively safe today, especially when compared to other energy conversion technologies and other risks, which we assume everyday without much thought. **Figure 6.10** shows the number of deaths in the five years from 2000-2005 in the United States from various common activities.[1] The nuclear industry is responsible for zero deaths during this time, while producing about 20% of US electricity. This is compared to over 30,000 deaths blamed directly on the coal industry during the same time. Iatrogenic deaths are simply due to natural causes, and contraception deaths are those associated with severe allergies to latex.

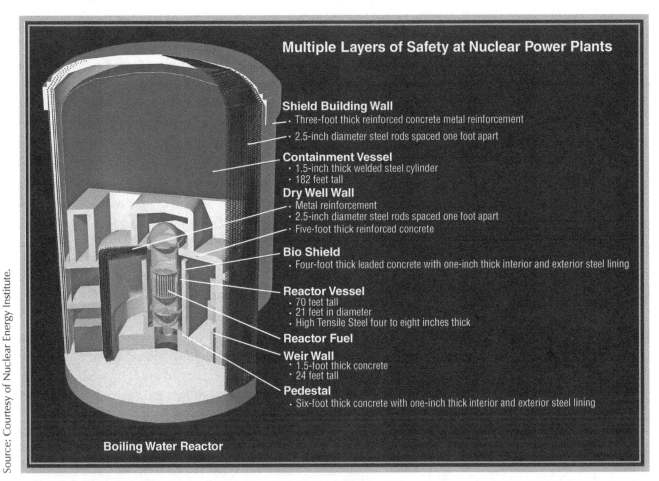

Source: Courtesy of Nuclear Energy Institute.

Multiple Layers of Safety at Nuclear Power Plants

Shield Building Wall
· Three-foot thick reinforced concrete metal reinforcement
· 2.5-inch diameter steel rods spaced one foot apart

Containment Vessel
· 1.5-inch thick welded steel cylinder
· 182 feet tall

Dry Well Wall
· Metal reinforcement
· 2.5-inch diameter steel rods spaced one foot apart
· Five-foot thick reinforced concrete

Bio Shield
· Four-foot thick leaded concrete with one-inch thick interior and exterior steel lining

Reactor Vessel
· 70 feet tall
· 21 feet in diameter
· High Tensile Steel four to eight inches thick

Reactor Fuel

Weir Wall
· 1.5-foot thick concrete
· 24 feet tall

Pedestal
· Six-foot thick concrete with one-inch thick interior and exterior steel lining

Boiling Water Reactor

This is not to say nuclear technology does not have its challenges. Nuclear experts, James Conca and Judith Wright, summarize the challenges with nuclear with five main issues: 1) cost; 2) risk; 3) waste; 4) mining; and, 5) fear—or public misunderstanding.[1] They (and many other nuclear experts) argue that the first four issues are solvable using modern technologies—both at the power plant and along the fuel cycle. The fifth issue, public misunderstanding, or fear, is substantially more complex, with human societies, economics, and political systems involved.

Currently, nuclear power provides about 8% of total US energy (~20% of US electricity), and is projected to remain a global leader in nuclear energy production for the next few decades. In fact, of the roughly 430 commercial nuclear reactors in operation in the world, the United States has 104. France is a distant second with 58 reactors; however, these provide about 80% of France's electricity simply because they consume drastically less electricity. Other nuclear giants include Japan, Russia, South Korea, and most European countries. Some countries are now phasing out of nuclear in light of cost, safety, and waste storage issues.

Figure 6.9 Layers of security within a modern nuclear fission reactor facility.

Activity	Number of Deaths in the U.S. in the past 5 years
iatrogenic	950,000
smoking	760,000
alcohol	500,000
automobile accidents	250,000
firearms	155,000
coal use (~50% of U.S. power)	30,000
construction	5,000
hunting	4,100
police work	800
contraception	750
nuclear Industry (~20% of U.S. power)	0

Source: Wright-Conca, 207

Figure 6.10 Deaths from various human activities in the US from 2000 to 2005.[1]

Chapter 6 Atoms and Atomic Energy

95

As with all other non-renewable energy conversion technology, nuclear fission produces wastes. Some of these are highly radioactive, and must be contained from the environment. A conventional fuel cycle for nuclear fission technology is presented in **Figure 6.11**. On the front end of the cycle, uranium-containing ore (<1% ^{235}U) is mined and milled, before being converted into a uranium-containing gas. The uranium-containing gas can then be enriched in a centrifuge process, which separates the heavier ^{238}U from the lighter ^{235}U fission fuel. The enrichment process is where the concentration of ^{235}U is determined; 3–5% for commercial-grade fuel and >90% for explosive/weapons-grade fuel. Once the concentration is established, the fuel is then processed into pellets, which are prepared for the nuclear reactors where they will operate for ~18 months, until it requires refueling.

After depletion, the spent fuel, along with various products from the fission, neutron capture, and decay reactions, is transferred to a storage facility to await final storage at a centralized site. The waste contains mostly low-level radioactive components, but the high-level waste takes more than 1,000 years to decay into stable isotopes, so it must be contained. In most countries, the spent fuel is reprocessed, where the unused ^{235}U, radioactive plutonium ^{239}P and other valuable isotopes (for research and medical applications, e.g., ^{60}Co) are recovered, with waste reduced substantially in volume. The US does not currently permit spent nuclear

Figure 6.11 Nuclear fuel cycle.

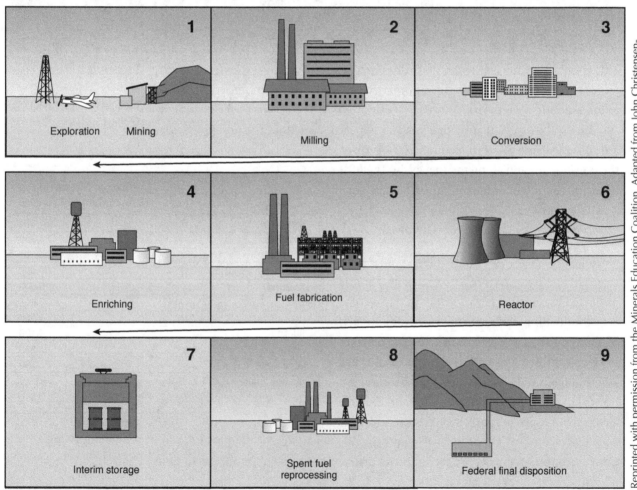

Reprinted with permission from the Minerals Education Coalition. Adapted from John Christensen-Teri Christensen, *Global Science: Earth/Environmental Systems Science, 7th Edition* (Dubuque, IA: Kendall Hunt Publishing Company, © 2009).

fuel reprocessing from commercial power plants. Final waste is typically stored in a central site, deemed safe from radiation leakage for well over 1,000 years. These centralized sites are currently under development in the United States, so most waste is stored in the interim on site in deep pools, or in small centralized sites awaiting final storage. More on nuclear fission energy and waste storage is at: www.nrc.gov/waste

Nuclear Fusion

Nuclear fusion is the combination of nuclei to make heavier nuclei. Fusion takes place in stars, including our sun, where enormous heat and gravity force atomic nuclei to combine to form heavier elements and release energy in the process. For example, in our sun, hydrogen nuclei combine with other hydrogen nuclei at a rate of about 4 million tons per second. This reaction produces helium nuclei and releases energetic neutrons and other radiation out into space. The basic fusion process in the sun is illustrated in **Figure 6.12**. In stars larger than our sun, atomic elements much heavier than helium are produced, and when these stars explode, they release their contents into the universe. Eventually, this matter collects into solar systems, including stars, planets, and moons. Ultimately, this is how every element on the periodic table, and all matter in the Earth System, including you and me, came to be.

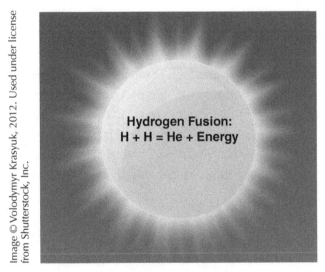

Image © Volodymyr Krasyuk, 2012. Used under license from Shutterstock, Inc.

Hydrogen Fusion:
H + H = He + Energy

Figure 6.12 Basic fusion processes within our sun.

Although fusion has been demonstrated in thermonuclear weapons (atomic bombs), it has eluded commercialization due to the challenging physics required in combining nuclei. In fact, the quote at the beginning of this chapter, "Our children will enjoy in their homes electrical energy too cheap to meter", from a 1954 speech by then US Atomic Energy Commission Chairman Lewis Strauss, was likely referring to nuclear fusion instead of fission, as it is now known that Chairman Strauss was supporting a secret nuclear fusion development project at the time. Therefore, controlled fusion for significant electric power generation on Earth has and may simply remain as a fundamental research endeavor for decades. There are, however, two international nuclear fusion projects worth exploring; 1) plasma-facilitated fusion, and 2) laser-facilitated fusion. In both cases, isotopes of hydrogen, deuterium (^{2}H or D) and tritium (^{3}H or T) are combined to make helium (^{4}He) and release energy and a neutron, as shown in **Figure 6.13**. This is often referred to as the D-T fusion reaction.

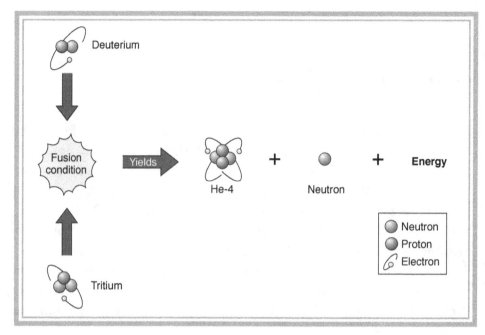

Reprinted with permission from the Minerals Education Coalition. From John Christensen-Teri Christensen, *Global Science: Earth/Environmental Systems Science, 7th Edition* (Dubuque, IA: Kendall Hunt Publishing Company, © 2009).

Figure 6.13
Deuterium-Tritium (D-T) fusion reaction.

Laser-facilitated (inertial containment) nuclear fusion technology is presented schematically in **Figure 6.14**. In laser-facilitated fusion, energetic photon beams (lasers) are focused onto a D-T containing pellet at the center of a spherical reaction chamber. As fusion takes place, heat is absorbed and transferred into water via a blanket system, making steam to power a turbine. Basically, the fusion reaction is used to generate thermal energy, boiling water into steam, which powers a turbine generator. A related technology is plasma-facilitated nuclear fusion, in which electromagnetic coils are used to magnetically-confine energetic D-T plasma in the center of a donut-shaped tube. As the D-T fusion takes place, a protective

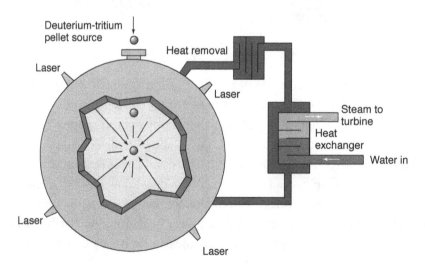

Reprinted with permission from the Minerals Education Coalition. From John Christensen-Teri Christensen, *Global Science: Earth/Environmental Systems Science, 7th Edition* (Dubuque, IA: Kendall Hunt Publishing Company, © 2009).

Figure 6.14 Basic features of laser-facilitated nuclear fusion reactor for steam generation.

blanket surrounding the plasma absorbs the heat, and transfers it into the water to make steam. More information on these technologies is at: www.lasers.llnl.gov/ and www.iter.org.

Both fusion systems are in various stages of development and demonstration in different locations around the world, and the principal operations have been successfully demonstrated for very short durations. It remains yet to be seen whether these designs can operate on a commercial-scale in terms of both time and power production. Both deuterium and tritium fuels are abundant; naturally-occurring deuterium is regularly extracted from water, and tritium can be produced through a nuclear fission reaction of lithium, an abundant element on Earth. As such, nuclear fusion is considered a renewable technology, and its continued research and development will likely yield important discoveries in physics, and ultimately a sustainable source of energy.

References

1. Wright, Judith, and James Conca. 2007. The GeoPolitics of Energy: Achieving a Just and Sustainable Energy Distribution: BookSurge Publishing, North Charleston, SC.

Quiz
··

(Open Book—Write Answers Below Numbers—Show All Work)

1. What are the three basic parts of the atom?

2. What is the difference between an ion and an isotope?

3. Photons comprise what form of energy?

4. How many neutrons are in oxygen-18 (^{18}O)?

5. What are the differences between nuclear fusion and fission?

6. What nuclear reaction drives commercial nuclear electric power plants?

7. What are the five issues with nuclear power?

8. How much of the US energy does nuclear fission provide?

9. What country has the most nuclear reactors in operation?

10. Do nuclear reactors require oxygen to operate?

11. What has more energy, radiation from Radio waves or X-Rays?

12. What are two nuclear fusion technologies under development?

13. When an electron jumps from one shell to another, what is emitted or absorbed?

14. Does nuclear fission release greenhouse gas?

15. What nuclear reaction powers our sun?

chapter 7

Renewable Energy Sources and Conversion Technologies

"To truly transform our economy, protect our security, and save our planet from the ravages of climate change, we need to ultimately make clean, renewable energy the profitable kind of energy."

— US President Barack Obama

Direct and Indirect Solar Power

On average, from about 93 million miles away, the sun provides Earth with about 1×10^{14} kW of continuous radiant power, emitted primarily in the visible spectrum with some ultra-violet (UV)—most of which is absorbed by stratospheric ozone. Over the course of a year (~8,760 hours) this continuous radiant power amounts to 1×10^{14} kW $\times$ 8,760 hours = 8.76×10^{17} kW-hrs of energy. This is the approximate total (100%) of annual solar energy input into the Earth System, as shown in **Figure 7.1**. Of this total, about a third (~31%) is reflected back into space by ice, snow, and clouds. About a fifth (~22%) evaporates water into clouds and drives the hydrologic cycle. Almost half (47%) is used to heat the oceans and land surfaces. A relatively tiny percentage (~0.2%) drives the winds and waves, and even less (~0.08%) is used in photosynthesis to support ecosystems. The degraded heat is emitted by the planet to cool itself off, primarily in the infra-red (IR) spectrum. Greenhouse gases (GHGs) in the lower atmosphere absorb some of this heat and warm the surface, increasing its average temperature.

On average, US citizens demand about 2 kW of total power, or use about 50 kW-hrs of energy per day for transportation, food, water, electricity, heating, cooling, etc. Even with current populations of ~7 billion, the sun supplies about 15,000 kW of power or about 350,000 kW-hrs of energy, per person every day. However, most of Earth's surface is covered by water, and the incoming solar energy on land is spread out over large areas, making it difficult to capture. Regardless, even if humans were able to convert just a tiny fraction of the total incoming solar energy into useable energy (e.g., electricity, transportation fuels), it would far exceed global energy demand.

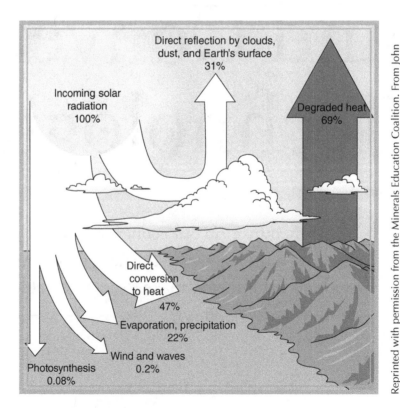

Figure 7.1 Solar power on earth and its distribution.

Reprinted with permission from the Minerals Education Coalition. From John Christensen-Teri Christensen, *Global Science: Earth/Environmental Systems Science, 7th Edition* (Dubuque, IA: Kendall Hunt Publishing Company, © 2009).

Solar energy is often distinguished into two types, *direct solar* and *indirect solar* energy. Direct solar energy is the energy directly converted into thermal energy or electricity by the sun's rays heating objects or powering solar-electric (photovoltaic) devices. Indirect solar energy is the subsequent conversion of the solar energy into biomass, wind, precipitation, and waves; in fact, all renewable energy sources, with the exception of deep, geothermal, are result (either direct or indirect) of solar energy.

Challenges in harnessing direct solar energy include: 1) most of it lands on and is absorbed by oceans; 2) it is diffuse and variable, mostly in low concentration across Earth's land surfaces; and, 3) it is only shining on about half of Earth at any given time. After being filtered through the atmosphere, the energy that reaches Earth's surface is regionally variable, from over 7 kW-hrs per m² every day, to as low as 3 kW-hrs per m² every day, on average. Since the average US citizen uses about 50 kW-hrs of total energy every day, using solar energy with an average 5 kW-hrs per square meter per day would require a 20% efficient solar-electric device with an area of about 50 m², or over 500 square feet per person. **Figure 7.2** shows a world map with average ground-level solar *insolation,* or energy per area per day (kW-hrs/m²/day) across Earth's land surfaces.

Figure 7.2 Average daily ground-level solar energy on Earth's land surfaces from 1983–2005.

Source: UNEP/GRID-Arendal.

Average annual ground solar energy (1983-2005)

7.5

7

6

5

4

3

Clear sky insolation incident, horizonthal surface (kWh/m²/day)

Solar Thermal

The most common use of solar energy is to directly convert it into thermal energy. Beyond warming the air, the oceans, the land and life, solar thermal energy can also be concentrated for cooking, heating water to warm buildings, and making steam to power turbines and electric power plants. One emerging application is the simple solar cooker, pictured in **Figure 7.3**. Whether in panel, parabolic or box

Figure 7.3 Example of a solar cooker—direct conversion of radiant into thermal energy (R → T).

configuration, the reflective surfaces of the solar cooker focus and concentrate the solar energy onto a central point, like a black cooking pot. As the pot absorbs the concentrated solar thermal energy, it heats the contents, e.g., boiling water or cooking food. A simple version of a solar cooker can be constructed from a cardboard box lined with aluminum foil. Several charitable organizations are making efforts to increase the use of solar cookers in developing countries; more information is at: www.solarcookers.org.

Solar thermal energy can also be used to heat water in buildings, as shown in the active closed-loop solar hot water heater in **Figure 7.4**. A thermal fluid, like antifreeze, is pumped through a flat plate solar thermal collector placed on the roof or external wall of a house or building, where it is heated by the sun. The hot fluid then transfers the heat into water inside heat exchanger. The hot water is then sent through the building for heating in radiation floor systems, or for hot water in cooking, cleaning, and hygiene. The system is considered active and closed-loop because the thermal fluid is actively pumped, and never leaves the system; it is continuously circulated depending upon the heat demand from the hot water heater. These systems can capture and transfer as much as 50% of the solar thermal energy into hot water. Where hot water use is prevalent, and solar energy is plentiful,

Reprinted with permission from the Minerals Education Coalition. From John Christensen-Teri Christensen, *Global Science: Earth/Environmental Systems Science*, *7th Edition* (Dubuque, IA: Kendall Hunt Publishing Company, © 2009).

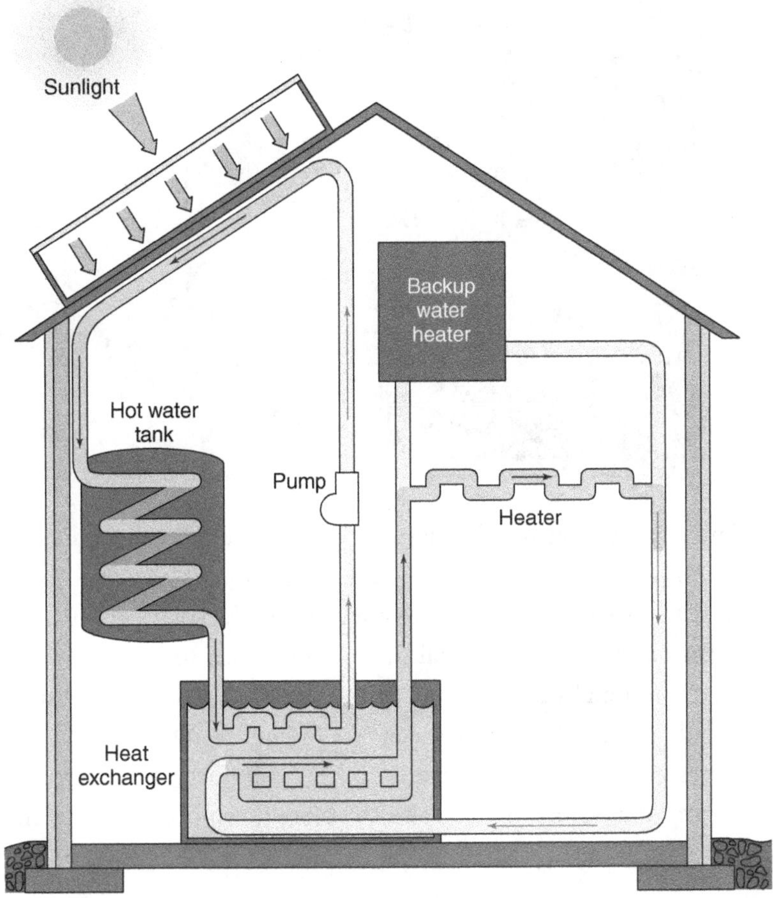

Figure 7.4 Active, closed-loop solar hot water heater.

solar-thermal hot water heaters can provide a reliable, low operating cost and low carbon heating alternative. Other, passive solar thermal systems utilize a large thermal mass (like stone or concrete) inside the building, which is exposed to a sun-facing window. As the mass warms, it retains and emits heat after the sun goes down. Additionally, other passive solar thermal systems, such as transpired solar collectors, are increasingly integrated into building HVAC systems.

Figure 7.5 presents a small array of direct solar thermal concentrator systems. These systems work by focusing the solar thermal energy onto a central point to power a Stirling Engine, which operates by exchanging a working fluid between a hot and cold source. The hot source is provided by the concentrated solar energy, and the cold source is provided by an air-exposed radiator on the backside of the engine. The working fluid drives a piston back and forth, which is connected to an electric generator. These devices can also follow, or track, the position of the sun in the sky to increase the collection efficiency. These small (~3 m diameter) *heliostatic* solar-concentrators have demonstrated peak electric power of over 3 kW, with more than 20% efficiency. This is comparable performance with commercial solar-electric systems and does not required water like steam-powered systems.

Figure 7.5
Concentrated solar thermal-electric conversion devices.

Other, larger-scale solar thermal systems include trough collector systems and solar *power towers*, as presented in **Figure 7.6**. The trough collector system (left) utilizes a field of reflective troughs to concentrate the solar energy onto pipes containing thermal fluid. The thermal fluid is then used to heat water into steam to power a conventional steam turbine connected to an electric generator. The power tower utilizes ground-based mirror arrays to concentrate the solar energy onto a central receiver tower, which houses a boiler, steam turbine, and electric generator. Some thermal fluids, like molten salts, can store thermal energy overnight, providing steady power output.

Figure 7.6   Large-scale, solar thermal electric power plants: ground-based troughs (left); solar power tower (right).

Solar Electric (Photovoltaic)

Solar electric devices, like the solar panels on rooftops or pocket calculators directly convert radiant energy into electricity (R > E). This works through the *photovoltaic* (PV) effect. The PV effect is when sunlight (electromagnetic radiation) ionizes certain materials and releases electrons. Using layers of different materials, PV cells (solar cells) direct the freed electrons into an electric current flowing from one layer into another and back again through an external circuit, which can power anything from a building to a pocket calculator. Basically, sunlight hits the PV cell and is converted into electricity, with about 10–20% efficiency in commercial systems.

Most commercial PV (solar) cells are made of ultra-purity (>99.9999%) silicon (Si), a semi-conductor material, meaning it isn't as electrically conductive as metal, but not an insulator like glass. The silicon is "doped" with other elements to make the PV layers. One layer of silicon is doped with an element having more electrons than silicon (to the right of Si on the periodic table). This creates an excess of electrons in the silicon layer, which is termed a negative or *n-type conductor*. The other silicon layer is doped with an element with less electrons than silicon (to the left of Si on the periodic table). This creates a deficiency of electrons in silicon layer, called *electron holes* or simply *holes*, and is termed a *positive* or *p-type conductor*. When the n- and p-type layers meet, a *p-n junction* is formed, as presented in **Figure 7.7**. Photons from the sun hit the p-n junction and mobilize electrons from one layer into the next, which are collected at contacts and transferred through an external circuit to the opposite layer. The electron flow through the external *load* is the electric power.

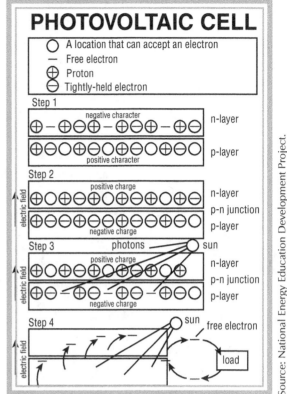

Figure 7.7 PV effect and electron flow in PV cells.

Source: National Energy Education Development Project.

PV *cells* are typically configured into bigger *modules*, which are then assembled into larger *arrays* of panels (**Figure 7.8**). The systems can be as small as mini-calculators (W-size), or large enough to cover roofs (kW-size); arrays can also be arranged to cover hundreds of acres in industrial-scale solar power plant systems (MW-size). The total installed capacity of solar PV is over 2,500 MW in the US, which ranks fifth behind Germany, Spain, Japan, and Italy.

Source: cell, module, array images: U.S. Department of Energy. Solar power plant image © Martin D. Vonka, 2012. Used under license from Shutterstock, Inc.

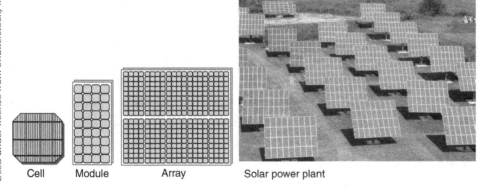

Cell   Module   Array   Solar power plant

Figure 7.8 Scale-up of PV from cells, to modules, arrays, and solar power plants.

The electric power produced in a PV cell is *direct current* (DC), meaning the electrons flow only in one direction. Most household electronic devices, including appliances, typically operate on *alternating current* (AC). Therefore, most PV systems are equipped with *inverters* to transform the DC into usable AC electric power. A residential-scale PV system is presented in **Figure 7.9**; this *grid-interconnected*, or *grid-tied* system requires an inverter to change the DC power from the PV array to AC power to be used in the house, or transferred into the utility grid. In *grid-independent*, or *off-grid* systems, batteries and back-up generators are often used in place of the grid. In both cases, the DC power from the arrays is inverted into AC, and then either used within the home, or either transferred back onto the grid (grid-tied) or stored in batteries (off-grid). When the sun isn't shining and the home demands electric power, the grid-tied system can simply switch to the electric utility service, whereas the off-grid system operates with batteries or the backup generator (typically burning fossil fuels), if continuous electric power is needed.

PV is becoming increasingly affordable with advances in materials and manufacturing, and government incentives. While economic payback is critical, so too is the recuperation of *embedded energy,* or the energy required to manufacture the PV array. Producing the ultra-pure Si required for the PV array is energy intensive, and it often requires years at full operating capacity to recuperate the embedded energy.

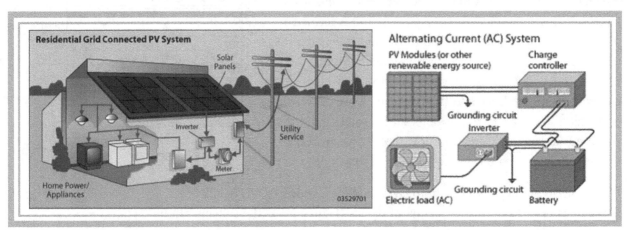

Source: U.S. Department of Energy.

Figure 7.9 Grid-connected residential PV system.

Biomass, Biogas and Biofuels

Biomass, biogas, and biofuels are all products of photosynthesis and subsequent processing through Earth's complex ecosystems. Biomass includes plants, animals, fungi, algae, bacteria, and all the organic (carbon-based) products, garbage, sewage, and other wastes they produce. Biomass is often distinguished into solids (biomass), liquids (biofuels), and gases (biogas). Biomass solids include wood-like matter, crop residues, and municipal solid waste (organic matter in garbage). Biogas includes products of biomass decomposition and flatulence from animals and bacteria, as they digest their respective foods. Biogas recovery and on-site burning for heat, steam and/or electricity has been in-place for decades at waste water treatment plants and various agricultural operations. Biofuels are liquids (primarily oils and alcohols) derived from plants, fungi, and algae, which can be subsequently processed into additives for gasoline and diesel. An increasing number of transportation vehicles are designed to operate on biofuels, ranging from pure (100%) biofuel, "B100", to a mix as low as 2% biofuel, "B2".

In general, biomass is used in place of, or in mix with, its fossil fuel counterparts (solid-coal, liquid-oil, and gas-natural gas). Therefore, as it is burned, it releases carbon dioxide and other pollutants. However, the carbon dioxide released in the burning is of similar amount to the carbon dioxide absorbed from the atmosphere during the original photosynthesis which produced the biomass. This is termed *net-neutral* in terms of GHG emissions.

Biomass is largely considered a renewable resource, but there are exceptions. In considering any renewable energy resource, we must consider the time that the resource regenerates itself vs. the time the resource has social utility. For example, if we harvest century-old trees that are needed for a biomass power plant, which operates for 25 years, it would take too long for the trees to regenerate to satisfy the power plants fuel needs. However, if the biomass fuel was fast-growing plants, algae, or fungi, such that the fuel can regenerate within days or months, the *renewability* of the biomass becomes more attractive.

The concept of biomass as part of a future, more-sustainable energy system is illustrated in **Figure 7.10**. Biomass, from conventional and new crops, along with

Figure 7.10 Biomass as integral part of regional energy system.

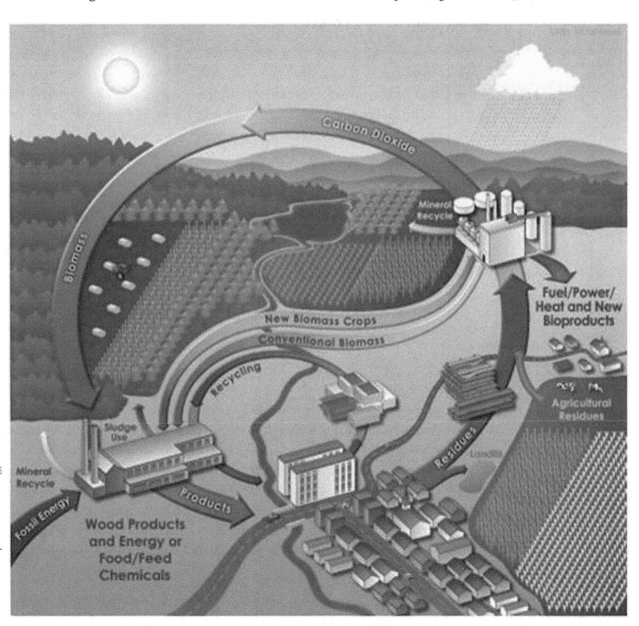

Source: US Department of Energy

recycled content, are fed with fossil fuels into electric energy and chemical production systems. These systems provide society with energy and products, with residues (wastes) either recycled, made into new products, or burned to recover the energy. The carbon dioxide emitted in the burning of the biomass and fossil fuels is then largely sequestered in the production of new biomass through photosynthesis.

Not all biomass is created equally. While wood and biogas are convenient for conventional boilers, furnaces, and turbines, the transportation sector currently operates on more than 90% liquid fossil fuels. Liquid biofuels, such as ethanol, have been touted as renewable solutions to this challenge. However, in the harvesting, energy-intensive processing, and distribution of the liquid biofuels, more carbon dioxide can be emitted than is sequestered, especially when deriving alcohol from low-sugar feedstock, such as corn. Additionally, some biofuels can divert crop land from food into cropland for fuels, spawning the so-called "food vs. fuel" concern. Additional concerns arise from the consumption of water, fertilizers, pesticides, and energy during the biofuel production.

One emerging biofuel crop is algae. Biodiesel can be derived from oils produced from photosynthetic algae. The algae can be grown in shallow ponds using brackish (salty) water and carbon dioxide. As the algae grow (consuming carbon dioxide and producing biomass), they are skimmed from the top of the ponds and harvested, extracting and converting the oil into engine-ready biodiesel. The algae can be grown in open or closed water systems, with the algae harvested and oils extracted and transported to a central *bio-refinery* to be processed into final product for distribution. The non-oil fraction of the algae can then be used as an amendment to fertilizer or animal feed. In the bio-refinery, processes such as gasification or pyrolysis may also be used to convert biomass into more-useful products. In gasification, the biomass is combined with steam to make a product biogas, which can be burned or used in making bio-based plastics. In pyrolysis, the biomass is heated in low oxygen environments, and separates into biogas, biofuels, and a carbon-rich solid residue called biochar, which is also used as a soil amendment.

Geothermal

On average, temperatures within Earth's crust increase by about 80°F for every mile beneath the surface. Geothermal energy conversion systems utilize the higher temperatures by extracting and transferring heat, or converting the heat into electric energy. Generally speaking, geothermal systems can be active or passive, or a combination of both. Active geothermal systems exchange a fluid with a geothermal source to heat the fluid. Passive geothermal systems include having a basement, or building into a cave or hillside. Both active and passive geothermal systems rely on the relatively constant temperatures within Earth's subsurface.

In a small-scale geothermal system, a heat pump is used to transfer heat into or from the building, depending on the season. For example, if the ground under a home held a constant temperature of around 60°F, this could be used for heating during the winter and cooling during the summer. An example of a small-scale geothermal energy system is shown in **Figure 7.11**.

Large-scale geothermal energy conversion systems take advantage of local hot-spots within Earth's crust, most near volcanic activity. In the United States, large-scale geothermal power plants are concentrated mostly in California, which in 2005 produced more electric power from geothermal sources than any nation in the world! Eight other states in the United States (Arkansas, Nevada, Utah, Oregon, Wyoming, Hawaii, Idaho, New Mexico) also have geothermal plants, but the power output is currently small in comparison. Total US installed capacity of geothermal is now over 3,000 MW (3 GW), by far the largest in the world. Iceland, well-known for its geothermal activity, ranks 7th, but their 600 MW installed capacity provides about 30% of their electricity, whereas the 3,000 MW in the United States supplies about 0.3% of our electricity.

In most geothermal systems, high-pressure steam is produced directly or indirectly in the geologic feature. Either way, the produced steam is fed into above-ground steam turbines connected to electric generators. An illustration and picture of a large-scale (MW-sized) geothermal facility is shown in **Figure 7.12**. Geothermal energy produces no carbon dioxide during operation, although emissions of volatile components within the geologic feature can occur. Appropriate site selection, and subsequent approval from residents and governments are currently the major obstacles for geothermal energy development.

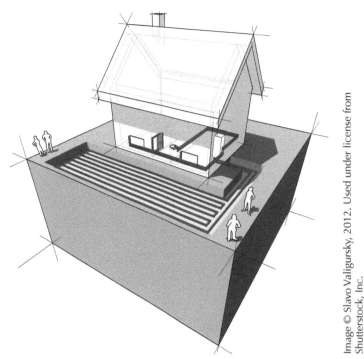

Figure 7.11 Example of small-scale geothermal energy system.

Figure 7.12 Large-scale geothermal electric power plant; illustration (left), image (right).

Energy from wind can be harnessed by converting the mechanical energy in the moving air (wind) into useable mechanical energy. Historically, wind has been used for transportation (sailing); it was then used for grain milling operations; and now, predominately for electricity production. Wind is essentially moving gas molecules, collectively making mechanical energy. When the wind hits turbine blades at the right angle, the turbine blades rotate about a central rotor connected to a generator. Modern wind turbines are typically comprised of cylindrical vertical shafts, some over 400 feet tall, with a generator hub at the top of the shaft connected to the rotor. This is primarily to take advantage of the relatively stable wind conditions higher off of the ground. The rotor connects to the turbine blades (typically 2 or 3 blades on modern turbines), which can span over 400 feet in diameter. These large, modern wind turbine systems can generate over 3 MW of electric power when operating in optimal wind conditions. However, some wind farms operate at maximum power only about 30% of the time, termed a *capacity factor* of 30, which accounts for downtime when wind isn't blowing or during regular maintenance. This means that if 100, 3 MW turbines (300 MW total) are installed at a site with a capacity factor of 30, the site can be expected to produce, on average, about 100 MW of electric power. **Figure 7.13** presents a diagram of internal wind turbine components (left)and a photograph of a modern wind turbine system (right).

Left: Reprinted with permission from the Minerals Education Coalition. From John Christensen-Teri Christensen, *Global Science: Earth/ Environemtnal Systems Science, 7th Edition* (Dubuque, IA: Kendall Hunt Publishing Company, © 2009). Right: image © dora modly-paris, 2012. Used under license from Shutterstock, Inc.

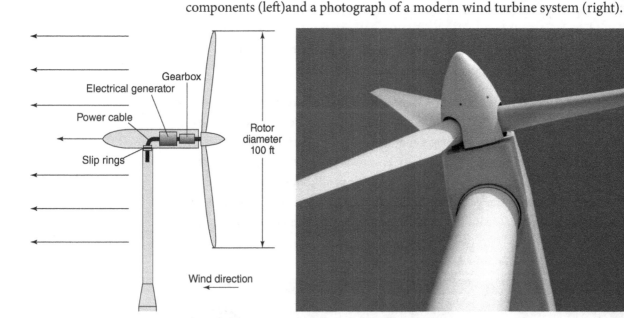

Figure 7.13 Modern wind turbine components (left) and image (right).

The power in the wind is determind by multiplying the mass flow rate of air blowing through the swept diameter of the blades (swept area) by the wind speed squared. The rate of air mass flowing through the turbine is also dependent on the wind speed. This means the wind power is directly related to wind speed cubed (Power = (wind speed)^3). This is why sites with steady, high-winds (like off-shore) are often preferred. Other wind energy technologies, such as tethered kites or sails driving ground-based energy conversion devices, are also being explored.

Hydro Power

Hydroelectric power plants (hydro power) convert mechanical energy in flowing water into electricity. This is similar to wind turbines, but water is about 1,000 times as dense as air, and its flow is more controllable and predictable through dams, etc. The amount of mechanical power available is determined by the mass flow rate of water and its vertical drop, which creates a head pressure pushing the water down. **Figure 7.14** presents a picture and schematic of a modern hydroelectric facility. Typically, the water is held back by a dam, which stores the mechanical energy of the water in a reservoir. When the water is allowed to flow through the penstocks within the dam, it drives turbines on the way to the outflow, typically a river below the dam. The turbine is connected to an electromechanical generator, which produces electricity to be distributed on the electric utility grid. Hydroelectric dams can also be used for *pump storage* by using electricity from intermittent wind and solar to pump water from below to above the dam for containment and use when needed through the turbines. Small-scale (<MW) hydropower devices, often referred to as "micro-hydro", are used in flowing rivers or streams to convert energy into usable mechanical or electrical energy on-site.

Figure 7.14 Hydroelectric power facility; diagram of common features (left), picture of Hoover dam (right).

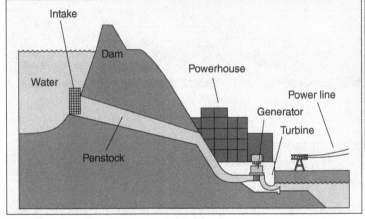

Left: © Kendall Hunt Publishing Company
Right: US Bureau of Reclamation.

Ocean Energy; Tides, Waves, and Gradients

Ocean energy includes the mechanical energy within moving tides and waves, as well as gradients in temperature and salt concentration (salinity), which can be converted into more useable forms of energy. **Figure 7.15** presents a simplified drawing for a tidal energy conversion system. Underwater turbines (connected to electric generators) turn as the water flows in and out with the tides. The process is also analogous to wind or hydro turbines, and flows are predictable.

Energy from oceans also can be harnessed from the mechanical energy in waves. **Figure 7.16** shows a wave energy conversion system under development. A floating buoy is tethered to the ocean floor; inside the buoy are a stationary electric coil and a magnetic shaft connected to the tether. As the buoy moves up and down, the magnetic shaft moves up and down within the electric coil, inducing an electric

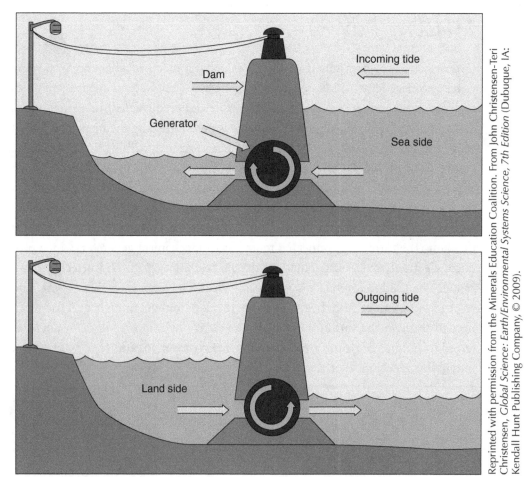

Reprinted with permission from the Minerals Education Coalition. From John Christensen-Teri Christensen, *Global Science: Earth/Environmental Systems Science, 7th Edition* (Dubuque, IA: Kendall Hunt Publishing Company, © 2009).

Figure 7.15 Tidal energy conversion system.

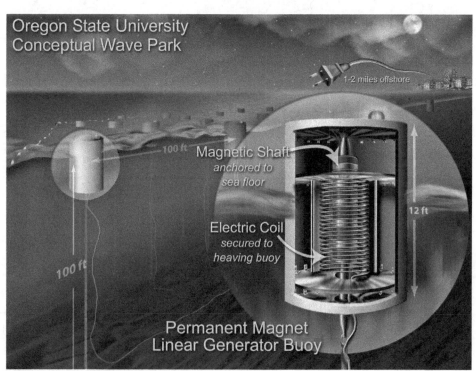

Source: Courtesy of Oregon State University.

Figure 7.16 Wave energy conversion system concept.

current and providing electric power to be transported on land. Another wave energy conversion system is one that funnels the waves into a box, which compresses air, making wind to drive a wind turbine within the box. The turbine is connected to an electromechanical generator to produce electric power.

Gradients within the oceans may also be harnessed to convert into usable energy. One of these concepts is the ocean thermal energy conversion (OTEC) system (**Figure 7.17**). The OTEC system takes advantage of the temperature differences between the top and bottom of the ocean, which can be over 90°F up top and less than 35°F near the bottom. If water of these temperatures are collected and fed into an OTEC system, the temperature gradient can be used to drive heat engines to produce electricity, desalinate water, and provide cold water for air conditioning and mariculture (fish farming).

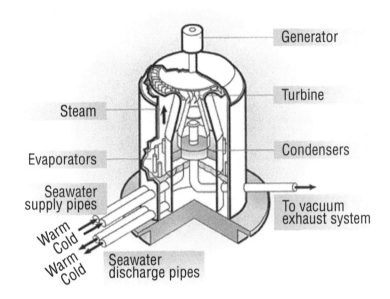

Figure 7.17 Ocean thermal energy conversion (OTEC) system.

Quíz

....

(Open Book—Write Answers Below Questions—Show All Work)

1. Approximately how much of the total solar energy input to Earth is used to heat land and oceans?

2. Approximately how much solar energy per area per day does your hometown get on average?

3. What are the uses for solar thermal energy?

4. What type of material is most commonly used in commercial PV panels?

5. What are the three types of biomass, and how are they used?

6. How much more power does wind have when its speed increases by a factor of 3 (e.g., 15 mph vs. 5 mph)?

7. How much geothermal power capacity does the United States have online?

8. Why are wind turbines so tall?

9. What two factors determine the amount of energy available for hydroelectric plants?

10. What are three ways to harness ocean energy?

chapter 8

Electricity and Hydrogen

"With distributed generation, every family, business, neighborhood, and community in the world is potentially both a producer and a consumer and vendor of its own hydrogen and electricity."

— Author Jeremy Rifkin, The Hydrogen Economy

Electricity Basics

Electricity is all around; from lightning bolts to the neurological activity in our brains. And while we marvel at the fascinating natural demonstrations of electricity, we also take it for granted in the countless electrical devices we associate with modern life. Electricity is defined as the flow of electric charge. Electric charge is either positive or negative, and its flow is facilitated by *charge carriers* in materials, such as protons (positive) and electrons (negative). The charge carriers flow through materials that are good *conductors* for the respective carriers, making for a net *current* of electric charge. Practically speaking, electricity largely involves the flow of electrons through metals (good electron conductors) and other materials.

The flow of electrons through a conductor is much like the flow of water through a pipe down a hill. The electric voltage, or electric potential (measured in Volts), is similar to the gravitational potential of water at the top of the hill. The electric current (measured in Amps) is similar to the mass flow rate of water through the pipe down the hill. When an electric circuit permits the flow of electrons, the electric power is determined by multiplying electric voltage by electric current, as described in **Equation 8.1**. Here, P represents electric power (Watts), I is current (Amps) and V is voltage (Volts). Similarly, the power of the water flowing through the pipe down the hill is determined by mass flow rate (kg/s) multiplied by the hill height (m) and by the acceleration due to gravity (m/s^2), which together makes for the SI units of Watts (1 W = 1 kg-m^2/s^3).

$$P = I \times V$$

Equation 8.1

Electrical devices, such as light bulbs, refrigerators and i-phones are considered electrical *loads* and create resistance to electron flow in order to convert the electrical energy into more-useful mechanical, thermal and/or, radiant energy. Electrical resistance is measured in units of Ohms (Ω) after the German physicist, George Ohm, who discovered that the electric current was directly proportional to the voltage applied. This direct relationship varied for different loads, but was often constant, or linear, in nature. Ohm determined the loads to have electrical resistance, much like the resistance flowing water encounters when flowing through a kink in a pipe. Ohm's law, described in **Equation 8.2**, shows the relationship of voltage (V) to the electrical current (I) and the resistance (R, measured in Ohms - Ω). With a little algebra, one can use equations 8.1 and 8.2 to determine the relationships among electric power, voltage, resistance, and current required for operation of any electrical device. Additionally, this information allows one to calculate electric energy costs of operating the device.

$$V = I \times R$$

Equation 8.2

Electric power is a measure of how much electric energy is used in any given time. Electric energy, therefore, is the electric power multiplied by the time that power is applied, as in **Equation 8.3**. Here, P is power (Watts - W), t is time (seconds - s) and E is energy (Watt-seconds, or W-s). In the water analogy, energy is similar to the total volume of water that flowed through the pipe and power is the flow rate any given moment. Although the basic unit of electric energy is the Watt-second (W-s), we use so much of it that it is more useful to measure it using a larger unit, the kilowatt-hour (kW-hr) — (1 kW-hr = 3,600,000 W-s). The kW-hr unit is the primary metric to monitor and charge for electric energy consumption around the world. Since energy units are interchangeable, the kW-hr has also emerged as a convenient means for comparing of all energy production consumption.

$$E = P \times t$$

Equation 8.3

Electricity comes in two practical forms; *direct current* and *alternating current*. Direct current (DC) is when electrons flow in one direction, and is produced in batteries, photovoltaics, and fuel cells and consumed in most computer electronics. In alternating current (AC), electrons flow back and forth—constantly vibrating in a wave of alternating positive and negative voltages, typically about 60 times every second (60 Hz) in the United States, and sometimes differently elsewhere around the world (depending upon electric grid design). AC power is often produced in large generators and can be easily transformed from high to low voltages in *transformers*, making it ammenable to transport long distances. Since AC evolved as the dominate electricity type on the utility grid in the United States, many compressors, motors, and lighting systems have been designed to operate on AC.

The debate between AC and DC distribution infrastructure continues, and newer, small-scale utility grid systems are taking advantage of some attributes of DC, e.g., allowing more photovoltaic input and avoiding conversion for electronic devices. AC and DC can be changed back and forth in devices known as *inverters*; however, each conversion results in some wasted heat. For example, the phone charger plugged into the wall outlet converts AC power into DC power to charge the DC-battery powered phone. If you feel the phone charger to be warm, it is to some extent due to the AC/DC conversion, which produces wasted heat in the process.

Electricity is often considered invisible, instant, and unstable. It is invisible, since we cannot see the electrons flowing through the conductors. It is instant, since the electric charge travels at incredible speeds (consumed almost same instant it is produced). And, it is unstable since it can fluctuate from moment to moment. The electric utility grid connects electricity producers with electricity consumers through a complex network of generation, transmission, and distribution. **Figure 8.1**

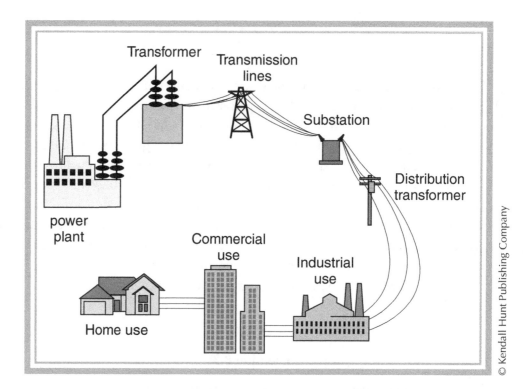

Figure 8.1 Simplified representations of electric power production, transmission and distribution.

illustrates a simplified network of centralized electric power production, transmission, distribution, and the three largest uses; industry, buildings, and homes.

At a typical power plant, fossil fuels are burned to boil water, making steam to turn a steam turbine connected to an electric generator. The electric power is of a specific voltage depending upon the generator design and operation. Typically, the voltage is increased, using transformers at a substation to create high-voltage for long-distance transmission. As shown in Equation 8.1, if the voltage is high, the current can be low for the same power. This allows for minimal losses during transmission, known as *line losses*. The high voltage electric power is then decreased using transformers at various substations to create voltages useful for industrial consumers (high voltages for heavy machinery) or lower voltages for businesses, residences, and other buildings.

The interconnectedness and magnitude of the electric utility grids in the United States and around the world can be seen in night images taken by satellite, presented in **Figure 8.2**. While the image does not show the full extent of the grid, it does present the lights resulting from its operation. A stable electric utility grid is now expected by citizens in most developed countries, whereas it is a luxury elsewhere around the planet. Access to electricity, poverty, and birthrate are all correlated globally. New, so-called "smart grid" technologies may help to alleviate the inefficiencies inherent in traditional electrical grid systems. How electric utility grids develop around the world will be interesting to observe.

Figure 8.2 Satellite image of the Earth at night.

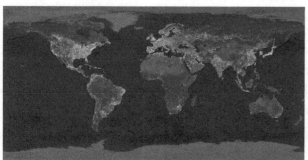

The percentage of electric power that comes from fossil fuels in any electric utility grid determines the amount of CO_2 emitted per kW-hr produced. Most often, emissions result from large (GW-size) centralized coal or gas burning power plants and distributed via the electric utility grid. The term *distributed generation* is now common to describe the combination of centralized power plants with distributed sources, such as wind turbines or photovoltaic systems. In the United States, the average GHG emissions are over one pound of CO_2 for every kW-hr produced, but this varies significantly from hydroelectric-rich northwest grid, to the coal-dominated southeast grid. Therefore, if an average US household used ~1,000 kW-hrs of electricity in one month, it would be responsible for ~1,000 lbs of CO_2 emitted into the atmosphere. This, along with fossil fuel consumption for providing food, clean water, heating and transportation define most *carbon footprints*. Numerous carbon footprint calculators are now available on the internet; one developed by the EPA for individual households can be found here: *www.epa.gov/climatechange/ emissions/ind_calculator.html*

The terms *energy conservation* and *energy efficiency* often get confused and misused. Energy conservation is simply not using energy, sometimes referred to as *nega(tive)watts*. Energy efficiency is getting the most from the energy that is used, and sometimes measured in miles per gallon or simply a percentage depending on what type of energy conversion is taking place (Figure 2.2). Of course, if you use energy more efficiently, you also avoid future energy use, or conserve it, so the terminology can get confusing.

The maximum electrical power consumed at any given moment, or the maximum electrical energy consumed during any one hour period in the billing cycle is considered the power *demand*. The demand refers to the increased demand on the utility infrastructure to deliver the needed electric power (generation, distribution, and transmission). The term *demand reduction* is also used frequently in energy discussions. Common electricity terms are described in **Table 8.1**.

Table 8.1 Common terms in electricity

Term	Description	Measure
Consumption	How much energy is consumed during billing cycle.	kW-hrs
Demand	Maximum energy consumed during a one hour period in billing cycle.	kW
Efficiency	How much we get out vs. what we put in	%, mpg
Conservation	Avoiding or minimizing energy consumption.	kW–hrs

The electric utility companies typically charge residential consumers by the amount of energy used during a billing period, whereas larger consumers pay both for energy and demand, to reflect the cost of the utility infrastructure needed to accommodate the demand. In some areas, utilities charge different amounts for electricity depending on when it is used; this is called "time-of-day" metering, which incentivize people and business to operate during low cost hours. In future *smart grid* systems, real-time energy charging could reflect the moment-by-moment fluctuations in the cost of producing electricity. Additionally, many utilities charge in a tiered system, in which consumers that use more energy than the average consumer pay a premium for that energy. Electricity consumers (large and small) can save money and reduce carbon emissions by simply conserving energy, being more efficient, and by reducing the maximum power that is drawn at any instant during a billing cycle.

Batteries and Fuel Cells

Batteries and fuel cells are both considered *electrochemical devices*; they convert chemical energy directly into electrical energy using electrochemical reactions (C > E). The difference between batteries and fuel cells is that batteries contain their fuel internally (stored chemical energy) and fuel cells are fed fuel from an external source (often called gas-powered batteries). In both cases, three main features of the electrochemical device are common; two electrodes separated by an electrolyte (dry or wet). One electrode is positive (called the cathode) and one electrode is negative (called the anode). Illustrated in **Figure 8.3**, anions (negatively charged ions) transport through the electrolyte to the anode, whereas cations (positively charged ions) transport through the electrolyte to the cathode. This tendency establishes an electric potential between the electrodes, which is directly related to the electrochemical reactions that take place on the electrode surfaces.

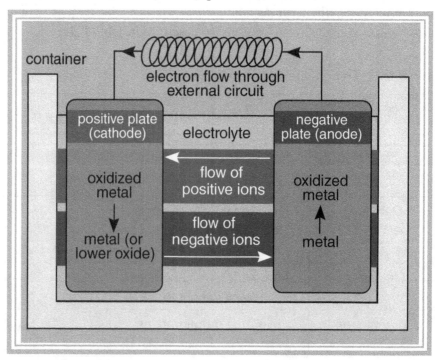

Figure 8.3 Basic components of a battery, or electrochemical cell.

Once an external circuit is connected, current flows both externally (via electrons) and internally (via ions) between the electrodes. At the anode, an *oxidation reaction* occurs wherein negatively charge ions from the electrolyte facilitate oxidation of metal or metal hydride into metal oxide or metal, respectively. The oxidation reaction liberates electrons to travel through an external circuit (powering a load), and to the cathode, where they drive a *reduction reaction*. The reduction reaction involves positive ions and electrons transforming metal oxide or metal into metal or metal hydride, respectively, and producing negative ions to continue the process. In the electrolyte, ions transport to each electrode to balance the loss or gain of electrons in the reactions, and there is an *internal resistance* to this transport specific to the ions, electrolyte, and electrochemical reactions.

Oxidation and reduction reactions are often referred to as *redox* (pronounced ree-dox) reactions. Redox reactions can be irreversible or reversible, depending upon the battery materials, and often batteries are categorized as primary (one-time use, or disposable) or secondary (rechargeable). The battery's electric power output characteristics are dependent upon the battery type and the load it is powering. Rechargeable battery types and characteristics are presented in **Table 8.2**. These include nickel-cadmium (NiCd), nickel-metal-hydride (NiMH), lead acid (common in cars), Lithium ion (Li-ion) and Alkaline batteries. Energy density describes how much energy is contained per mass of the battery, measured in Watt-hour per kilogram (W-h/kg). For reference, a 100 Watt light bulb on for one hour uses 100 W-hrs of energy. The charge time and cell voltage are a result of the redox chemistry, which is unique to the electrode and electrolyte materials used.

Industrial-scale battery manufacturing is now completely automated, and involves numerous relatively simple steps to construct a common cylindrical battery, like a size AA. Automated industrial processes are now employed worldwide to create batteries with voltages, load capacities, and geometries tailored specifically for the application they are intended to power. Some large-scale (>MW), flow cell batteries are being used to back-up intermittent solar and wind-powered electric generation systems, providing steady, more reliable electricity (better power quality) to the grid.

Table 8.2   Common Rechargeable (secondary) Battery Types and Characteristics

Battery Types Approximate Values	NiCd	NiMH	Lead Acid	Li-ion	Reusable Alkaline
Energy Density (W-h/kg)	45-80	60-120	30-50	110-160	80
Charge Time (hours)	1	2-4	8-16	2-4	2-3
Cell Voltage (volts)	1.25	1.25	2	3.6	1.5

Fuel cells operate in much the same way as batteries, with both oxidation and reduction reactions occuring at electrodes, and ions transporting through the electrolyte. In most cases, the fuel for the fuel cell is hydrogen gas (H_2) and the oxidant is oxygen (O_2), typically from air. As with batteries, there are several different types of fuel cells, depending upon the materials used. **Figure 8.4** presents various common fuel cell types and their characteristics. Beginning at the top are the higher temperature fuel cells, which can process (or reform) hydrocarbon fuels internally —making hydrogen (H_2) and carbon monoxide (CO) fuel from conventional hydrocarbon fuels. At the top of this list is the solid oxide fuel cell (SOFC), which operate between about 500–1000°C, and ultilize solid-state electrolyte through which negatively charged oxygen ions (O^{-2}) transport. Next is the molten carbonate fuel cell (MCFC), which operate at about 650°C and utilize a molten carbonate electrolyte through which negatively charged carbonate ions (CO_3^{-2})transport. As such, the MCFC has a unique requirement of carbon dioxide (CO_2) to be fed with the air to facilitate carbonate ion transport in the electrolyte.

Lower temperature fuel cells require external fuel processing to refine hydrocarbon fuels into useable hydrogen, which needs to be relatively pure—with carbon monoxide (CO) and other trace gases removed. The first on this list is the phosphoric acid fuel cell (PAFC), which operate at about 200°C using an acid electrolyte to transport protons (H^+). The proton exchange membrane fuel cell (PEMFC) operates at about 80°C using an acidic polymer membrane electrolyte, through which positively charged protons (H^+) transport. The PEMFC is the most common type

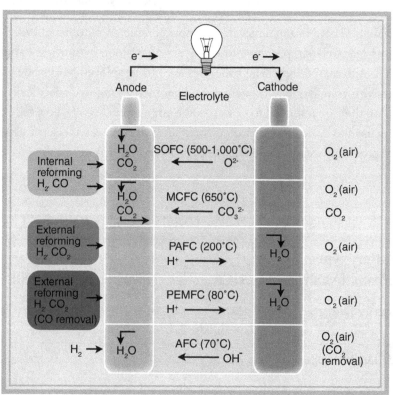

Figure 8.4 Fuel cell types and operation characteristics.

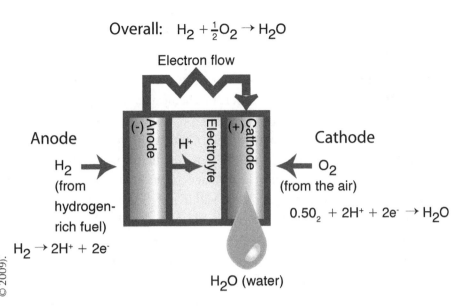

Overall: $H_2 + \frac{1}{2}O_2 \rightarrow H_2O$

Electron flow

Anode

(-) Anode | Electrolyte | (+) Cathode

H^+

Cathode

H_2 (from hydrogen-rich fuel)

O_2 (from the air)

$0.5O_2 + 2H^+ + 2e^- \rightarrow H_2O$

$H_2 \rightarrow 2H^+ + 2e^-$

H_2O (water)

Figure 8.5 PEMFC operation basics.

for vehicle and portable applications. Finally, the alkaline fuel cell (AFC) operates at about 70°C and utilises a basic liquid solution to transpot negatively charged hydroxide ions (OH^-). The AFC type was used in the Apollo missions to the moon to produce electric power and drinking water for the astronauts.

Figure 8.5 further illustrates the basic operation of a PEMFC; often simply called a hydrogen fuel cell. Hydrogen molecules (H_2) are fed into the anode, where they break apart in the presence of a *catalyst* (usually platinum) into 2 protons (H^+) and 2 electrons (e-), in the reduction reaction shown in **Equation 8.4**. This is called the *half-cell reaction* for the anode.

$$H_2 = 2H^+ + 2e^-$$

Equation 8.4

The electrons transport through an external cicuit to power a load, like a light bulb. The protons transport through the polymer electrolyte to the cathode, where they meet the electrons and oxygen molecules (O_2) to produce water (H_2O). This cathode-side half-cell oxidation reaction is presented in **Equation 8.5**.

$$0.5O_2 + 2H^+ + 2e^- = H_2O$$

Equation 8.5

Combining the two half cell reactions, we get the overall cell reaction in **Equation 8.6**. This is identical to the combustion reaction and releases the same amount of energy; however, the fuel cell combines the fuel and oxidant through an electrochemical process to convert chemical energy directly into electrical energy (C > E), compared to converting chemical energy into thermal energy (C > T) in combustion.

$$H_2 + 0.5O_2 = H_2O$$

Equation 8.6

Electric Generators

At the heart of most electric power plants in the world are *electromechanical generators*, or electric generators, which convert mechanical energy into electrical energy (M → E). Generators capitalize on the electromagnetic principle that all electric current is accompanied by a magnetic field. Conversely, the movement of a magnetic field induces an electric current within conductive materials, for example, copper. **Figure 8.6** illustrates this electromagnetic effect and how electric current can be produced by moving a conductor within a stationary magnetic field. Generators range in size from hand crank size (W) to portable size (kW) to power plant size (MW). The mechanical input to a power plant generator typically comes from a steam turbine shaft, powered by steam from a thermal source (T → M), usually fossil fuel combustion or nuclear fission reactions (C → T or N → T), and sparingly but increasingly from concentrated solar and geothermal sources (R → T). Mechanical input to generators can also come from turbines powered by the wind and flowing water. Most electromechanical energy conversion devices are now very efficient (>90%), and built with increasingly sophisticated materials and designs.

Figure 8.6 Basics of electromechanical generator operation.

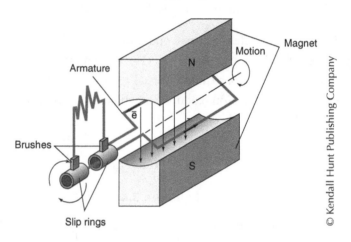

© Kendall Hunt Publishing Company

Motors

Electric motors, with a wide-range of sizes, operate everywhere around us—they run the fans in our ventilation systems (W) compressors in our refrigerators (kW), and the drive systems on train locomotives (MW). Electric motors are essentially the reverse of the electric generator; instead of converting mechanical into electrical energy (M → E), motors convert electrical energy into mechanical energy (E → M). The mechanical energy output from a motor is typically in the form of a rotating shaft, which can then be converted into other useful mechanical energy through gears and linkages. The basic features of an electric motor are presented in **Figure 8.7**. The central rotor shaft is electromagnetic, which spins when electricity flows through the electric coils surrounding it. Like modern generators, modern electric motors are also very efficient (>90%) and are being designed with increasing sophistication and less expensive materials. Small, eccentric motors

Electrical Energy Input Spins Electromagnetic Coil Rotor,
Which is Connected to Drive Shaft (Mechanical Energy Output)

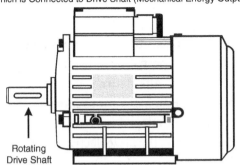

Rotating
Drive Shaft

Figure 8.7 Basic features of an electric motor.

create regular mechanical vibrations with high-speed wobbling shaft rotation, and are now common in toothbrushes, shavers, and other vibrating electronic devices.

Thermoelectric Systems

Another type of energy conversion, which can either generate electricity directly from a thermal energy gradient (temperature difference), or consume electricity to produce a temperature gradient (for heating or cooling), is *thermoelectric* technology. Thermoelectric modules are similar to PV modules, and employ semiconductor materials with p- and n-type conductivities. The basic design and operation of thermoelectric modules are shown in **Figure 8.8**. If an electric current is applied by an electric power source, a thermal gradient is produced (Figure 8.8 left) (E→ T). Termed the *Peltier Effect*, after the French physicist who discovered it in 1834, thermoelectric cooling is now used in most small (dormitory-sized) refrigerators and laptop computers.

Alternatively, if a thermal gradient is applied (a hot source and cold source) an electric current is produced (Figure 8.8 right) (T → E). This is termed the *Seebeck Effect* for the German physicist who discovered it in 1821. *Thermoelectric power generators* are used frequently in satellite operations using radioactive thermal sources, called *radioisotope thermoelectric generators* (RTGs). Other applications include electronics, such as wrist watches powered by body heat, and larger-scale systems to convert waste thermal energy from exhaust systems into useable electricity. Additionally, the effect is used in thermocouples—a common device for measuring temperature.

Figure 8.8 Basic design and operation of thermoelectric devices, heating/cooling (left), power production (right).

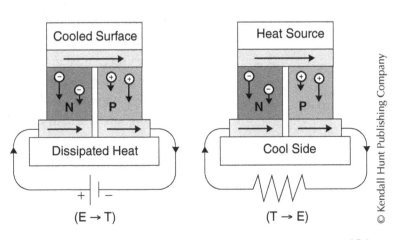

Piezoelectric Systems

One way to convert mechanical energy to and from electrical energy is by *piezoelectric materials*, which have a unique property of producing an electric voltage if they are squeezed or stretched (M → E). Alternatively, if electric voltage is applied, the material expands or contracts (E → M). Various innovative applications of piezoelectric materials also have been developed, such as in club dance floors. As the people dance on a spring-loaded floor, the piezoelectric materials within the floor produce a small amount of electric power (M → E), which is used to provide lighting through colorful Light Emitting Diodes (LEDs). The basic design and operation of piezoelectric materials are shown in **Figure 8.9**.

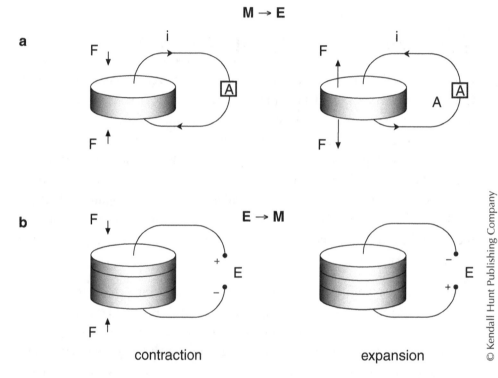

Figure 8.9
Piezoelectric materials and characteristics—electric voltage generation by squeezing or stretching (a), converting voltage to mechanical action (b).

© Kendall Hunt Publishing Company

Hydrogen

Hydrogen is an attractive means for storing chemical energy since its combustion (either directly or electrochemically through fuel cell) produces only water. However, most of the hydrogen used in the world today is derived by methane (CH_4) in natural gas, a non-renewable fossil fuel. Typically, water (H_2O) is combined with methane (CH_4) to create hydrogen and carbon dioxide (CO_2) in an endothermic (energy-requiring) reaction, shown in **Equation 8.7**. This reaction not only produces GHG, it also considered a non-renewable means of hydrogen production, since it is derived from fossil fuel and requires heat (typically from fossil fuel combustion).

$$CH_4 + 2H_2O = 3H_2 + CO_2$$

Equation 8.7

Non-renewable hydrogen from natural gas, with carbon capture and storage (CCS), has been proposed as a transitional strategy into a more-sustainable renewable hydrogen-powered economy. In this system, natural gas would be fed into electric power plants and hydrogen production plants, with the carbon dioxide separated and sequestered. The electricity and hydrogen would then be distributed to end users for electric demands, heating, and transportation—the users could either burn the hydrogen like natural gas in furnaces or turbines (producing thermal or mechanical energy), or in a fuel cell (generating electricity)—both emitting water in the process.

Renewable hydrogen can be produced using electricity from renewable sources and *electrolysis*, or the splitting of water using electricity. The basic concept of electrolysis is presented in **Figure 8.10**. As electric current is driven through water (H_2O), the reverse of combustion occurs—it separates into hydrogen (H_2) and oxygen (O_2), which bubble out of the water. If the electricity is produced by renewable sources, such as PV, wind or hydro, it is considered *renewable hydrogen*.

Figure 8.10 Basics of electrolysis—"splitting water" using electricity.

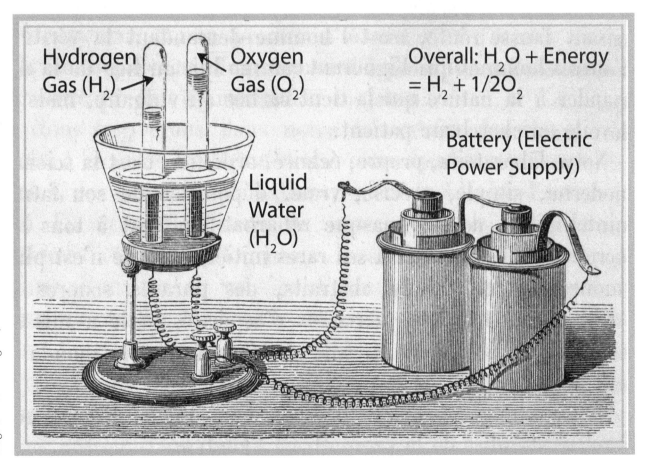

A hydrogen-based energy system may also facilitate integration and complement renewable energy sources, which are often intermittent (wind not always blowing, sun not always shining, and biomass not always growing). Through electrolysis, renewable electric energy can be converted into hydrogen and stored for use in fuel cells to produce electricity on demand or through combustion-based hydrogen vehicles. Additionally, hydrogen can be derived from biomass through pyrolysis and added to the distribution system. This interconnected renewable hydrogen energy system is illustrated in **Figure 8.11**. While a transition to this system may require decades to develop, it has emerged as a promising sustainable energy strategy.

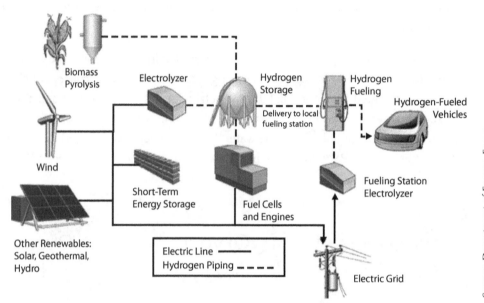

Figure 8.11

Renewable hydrogen-based energy system for electricity and transportation.

Source: Department of Energy: Energy Efficiency and Renewable Energy.

Quiz and Problems
....
(Open Book—Write Answers Below Questions—Show All Work)

Quiz
Answers in Spaces and Show All Work:

1. If an electric power supply operated at 10 A and 5 V, how much electric power is available?

2. If an electric device had a rated resistance of 10 Ω and the wall voltage was 120 V, how much current *and* power does the device consume?

3. If a 60 W light bulb operates for 1,000 hours, how much energy does it use?

4. Do batteries produce AC or DC power?

5. What is the process called for splitting water using electricity?

Problems

Show all work:

1. Calculate and report your individual or family's carbon footprint using this calculator: www.epa.gov/climatechange/emissions/ind_calculator.html

2. Suppose a family of four is considering purchasing a solar-electric device for their home. Annually, they currently consume ~10,000 kW-hrs of electricity and pay $0.10 / kW-hr. Assume a 2.5 kW solar-electric device operates perfectly for 4,000 hours a year and costs $20,000 (installed). Approximately how many years will it take to break even on this investment without any incentives or rebates?

3. Suppose you have a total of five (5) "vampire loads" in your house, each of which requires 10 W of power as long as they are plugged in (examples of "vampire loads" are cell phone chargers and plasma TVs on standby—always on, "sucking" electric power). If operated every day for one year, calculate the annual cost of electricity to power these loads if charged $0.10 / kW-hr.

chapter 9

Food and Water

"When the well is dry, we know the worth of water."

— Benjamin Franklin

Food

Food provides the energy and nutrients that sustain life. More accurately, metabolizing the foods we eat along with oxygen from the air we breathe and water we drink, converts stored chemical energy into usable energy within our bodies, while replenishing essential vitamins and minerals. Food comes in a variety of forms and flavors around the world. Major food groups include those derived from plants, such as fruits, vegetables, nuts, and grains, and those derived from animals, such as dairy products, meats, fish, and insects, which mostly feed on plant-based diets. Within these foods are carbohydrates, proteins, and fats, which are metabolized for energy, along with various vitamins and minerals, required for healthy living. **Table 9.1** provides examples and a breakdown of the major food groups. In addition to these major food groups, many modern (processed) foods also contain: artificial flavors, including salts and sweeteners; additives, such as colors, thickeners, and preservatives; and, some foods are subjected to electromagnetic radiation treatments to prevent various food-born diseases.

The stored energy content within food is often measured in kilocalories or *Calories* (with a capital C). Daily caloric intake varies drastically around the world. **Figure 9.1** presents the average, per-capita daily caloric intake in various countries around the world, as prepared by the United Nations Food and Agriculture Organization (FAO). People in some developed countries consume more than 3,500

Table 9.1 Examples of Major Food Groups and Their Contents

Derived from:	Food Group	Primary Contents	Examples
Plants	Grains	Carbohydrates	Whole Grains (Oatmeal, Brown Rice, Millet), Refined Grains (Corn or Flour Tortillas, White Rice/Bread)
	Fruits	Carbohydrates	Bananas, Papayas, Berries, Oranges, 100% Fruit Juices
	Vegetables	Carbohydrates	Broccoli, Carrots, Corn, Asparagus
	Nuts and Seeds	Proteins/Fats	Peanuts, Pine Nuts, Almonds, Cashews, Walnuts, Sunflower Seeds
	Legumes	Proteins/Fats	Black Beans, Pinto Beans, Soybeans, Peas
Animals	Dairy	Proteins/Fats	Milk, Cheese, Yogurt
	Poultry	Proteins/Fats	Turkey, Chicken, Duck, Eggs
	Meats	Proteins/Fats	Beef, Pork, Lamb, Wild Game
	Fish and Seafood	Proteins/Fats	Codd, Salmon, Tuna, Crab, Shrimp, Lobster

Calories per day, whereas people in some developing countries consume as little as 1,500 Calories per day, or often much less. In both diets, excess or deficiency of certain nutrients causes malnutrition, which the World Health Organization rates as the single largest threat to human health. And while there are sufficient food resources to feed the population, *undernourishment* (starvation/hunger) continues throughout the world. In fact, it is estimated that about fourteen percent of total global population, or almost one billion people, are undernourished today. At the same time, many developed countries, including the United States, face epidemic obesity, which drastically increases adverse and costly health issues.

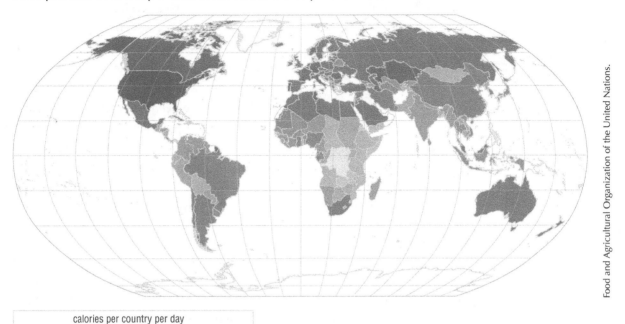

calories per country per day

| 1,900 | 2,300 | 2,700 | 3,100 | 3,500 |

| 1,700 | 2,100 | 2,500 | 2,900 | 3,300 | 3,700 |

Food and Agricultural Organization of the United Nations.

Figure 9.1 Average Daily Caloric Intake by Country

Food Systems

Historically, humans relied on hunting and gathering for food resources and diets consisted primarily of wild plants and animals. After the agricultural revolution, basic farming and livestock operations evolved, and diets grew into domesticated "cereal" grains and meats. With the Industrial Revolution, fossil fuel-powered engines largely replaced human and animal labor in the fields and food processing facilities, and slowly transformed agriculture into the globally-interconnected *food-systems* of today. Diets are now highly-variable around the world, and foods range from raw to highly-processed. As presented in **Figure 9.2**, the basic components of most modern *food systems* include:

- resources—such as land (soil), machinery, and labor force;
- production—where food is cultivated and harvested;
- processing—where food is transformed and packaged into retail products;
- eating—where we consume the food;

Patricia Allen, Jan Perez, Phil Howard and Elliott Kuhn (graphic artist), 2004. University of California, Santa Cruz

The Food System

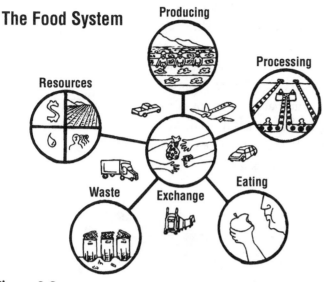

Figure 9.2 Basic Components of a Food System

- waste—scraps from eating, production or processing; and,

- exchange—where products are distributed and transactions are made.

The exchange process is necessary to facilitate food systems, and is often governed by standards, regulations, and other policies. It is the means by which value (like money) is transferred in exchange for products or services. In modern agriculture, exchange processes and supply-chains are often at a global-scale. For example, the bananas in your local supermarket may have come from South America, whereas the gourmet cheese down the aisle may have been produced in France. Before the 1950s, more-localized food systems were prevalent in the United States; however, today the majority of food-systems are regional, national, or even international. Although highly-productive, global food systems are arguably more-vulnerable compared to localized food systems, which consume substantially-less energy resources for transportation, processing and fertilizer production.

Agriculture is dependent upon *soil,* which is a complex and variable mixture of air, water, minerals, and organic matter. Soil is the connection between Earth's living (biosphere) and non-living systems (lithosphere, hydrosphere, and atmosphere) by providing the nutrients required for healthy plant and animal growth. The scientific study of soil is called *pedology.* Different *soil types* include: *clay; sand;* and, *silt.* Clay is sticky and holds water tightly. Sand is porous and lets water flow right through. Silt is smooth and slick and holds water reasonably well. Loam is a mixture between clay, sand, and silt, and is smooth and slick, but somewhat sticky and loose. Typically, soils are some ratio between clay, sand and silt, mixed with organic component, called *humus,* along with water and air content.

Many soils support rich ecosystems, both above and below ground, e.g., micro-organisms and burrowing animals. Above ground, plants growing in the soil require sunlight, carbon dioxide (CO_2), water (H_2O) and nutrients; primarily the six *macronutrients,* nitrogen (N) phosphorous (P) and potassium (K), and to a lesser-extent calcium (Ca), magnesium (Mg) and sulfur (S). Additionally, at least seven *micronutrients,* such as boron (B), copper (Cu), iron (Fe), chlorine (Cl), manganese (Mn), molybdenum (Mo), and zinc (Zn) enhance plant growth when present in trace amounts. Soils naturally transfer these nutrients from the lithosphere, atmosphere, and hydrosphere to *fertilize* the plants via their root systems.

Since the Industrial Revolution, synthetic (human-made) fertilizers from raw materials, including mined phosphate and potash ores (containing phosphorous and potassium), along with ammonia (NH_3)—produced from the combination of nitrogen (N_2) from air and hydrogen (H_2) from natural gas (CH_4), have substantially increased crop yields and now support over half of the global food system. Domesticated animals feed off of plants and require similar nutrients for optimal health. Animal feed manufacturing and concentrated animal feed operations (CAFOs) have also accelerated food production. In the United States, CAFOs are regulated by the EPA, since they generate industrial-scale emissions of both water and air pollutants, as well as greenhouse gases (from animals and their wastes). Animal treatment (husbandry) has also transformed substantially with the development of CAFOs, often necessitating medications to maintain animal health and optimize productivity.

Modern agriculture consumes large amounts of energy and water, making these resources and their constraints interdependent. This is sometimes referred to as the "energy-water-food nexus." Additional challenges of modern agriculture include the erosion and nutrient-depletion of soils with plowing and the cultivation of single crops (monocultures), which are vulnerable to pest infestations and diseases. In spite of these challenges, the development of synthetic fertilizers, herbicides, pesticides, genetically-engineered crops, and mechanized production have demonstrated the potential to feed the global population. However, access to these technologies or food resources is not uniform around the world, and many people continue to die of starvation every day.

Food prices are often used an indicator of stability within the globalized food system. People in developed countries pay only a small fraction of income on food, whereas undernourished people in developing countries pay a much higher percentage of income for food and are therefore particularly sensitive to price changes. **Figure 9.3** presents the FAO global average food price index, with an index of 100 indicating average prices during the period from 2002 to 2004. Due to the interconnected global food-system, extreme weather events, such as regional droughts and floods, as well as regulatory and political systems can have significant impacts on global food prices, which have been especially volatile in the last few years.

Figure 9.3 FAO Food Price Index

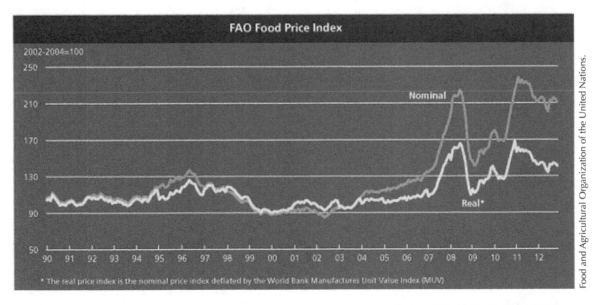

Food and Agricultural Organization of the United Nations.

Water

Water (H_2O) is referred to as many things; including the universal solvent, the ultimate purifier and life's blood. It is essential to life as we know it on Earth. Pure water (H_2O) consists of two hydrogen atoms bonded to one oxygen atom, and exists in three phases; solid, liquid, and vapor. Water is distributed throughout Earth's hydrosphere in the oceans, snow and ice, and freshwater, as presented in **Figure 9.4**. Water moves throughout the hydrosphere driven by evaporation from the sun in the hydrologic cycle (Figure 2.3). Most (97%) of Earth's water resides in oceans, where it has dissolved minerals over time, making it salty, or *saline*. Of the 3% freshwater remaining, most of it (about 69%) is retained in icecaps and glaciers where much of it has been for thousands of years. About 30% of the Earth's freshwater exists as *groundwater*, which fills the pore-space between rocks underground. Groundwater includes potable (drinkable) *aquifers*, which are often exploited using wells for above-ground water supplies. The remaining water, about 0.03% of the total, is either in the atmosphere or on land surfaces, including rivers, swamps, and lakes. Most humans utilize this *surface water* on account of easy access. Unfortunately, both ground and surface waters are also subject to contamination from human activities.

Circulation within oceans plays a vital role in Earth's climate, both on the surface and in the deep. On the surface, the sun warms the land and oceans differently, causing waves and currents pushing warm and cold water around the planet. As

Figure 9.4 Distribution of Water in Earth's Hydrosphere

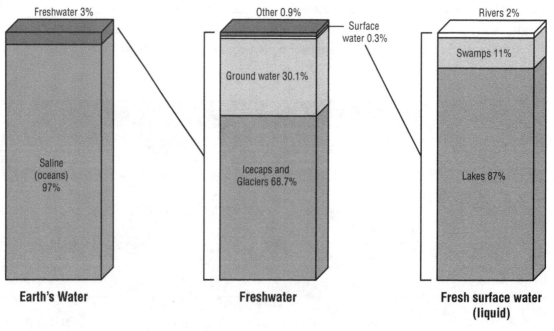

Distribution of Earth's Water

Source: http://ga.water.usgs.gov/edu/waterdistribution.html

warmer water approaches colder regions, it freezes into ice on the surface, which pushes salt into the deeper water. The salty water is denser than freshwater, and sinks deep within the water column. This process drives a global ocean circulation pattern, often referred to as the great ocean conveyor, or the *thermohaline circulation*, from differences in temperature (thermo) and salt concentration (haline). **Figure 9.5** presents the global ocean thermohaline circulation. This circulation plays a pivotal role in regulating Earth's climate, by absorbing and transporting thermal energy throughout the Earth's System. Recent and dramatically increasing CO_2 levels in the atmosphere have had significant impacts on ocean circulation (temperature changes), as well as ocean volume (sea level) and its composition (ocean acidification). These impacts affect both marine and terrestrial climates and challenge the complex ecosystems they support.

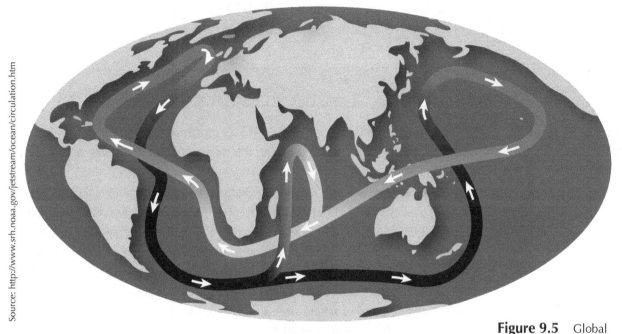

Source: http://www.srh.noaa.gov/jetstream/ocean/circulation.htm

Figure 9.5 Global Ocean Circulation Patterns

Behind oceans, ground water is the largest reserve of water in the hydrologic cycle. Ground water exists in saturated regions in the subsurface, below the water table, which varies in depth around the world. **Figure 9.6** illustrates the basic features of ground waters and their flows between aquifers and surface waters. The flows into and out of ground water are termed *recharge* and *discharge,* respectively. As surface waters and precipitation percolate through the ground, it recharges the ground water and can raise the *water table.* As we pump water from the ground, or as it flows into surface water, it discharges the ground water and can lower the water table. Some ground water is relatively isolated in confined aquifers, where it can remain for thousands of years. When tapped, these aquifers are referred to as 'artesian wells,' which sometimes flow to the surface under the pressure within the aquifer. Currently, withdraws from ground water sources provide about 25% of the water used in the United States for agriculture, industry and drinking/hygiene.

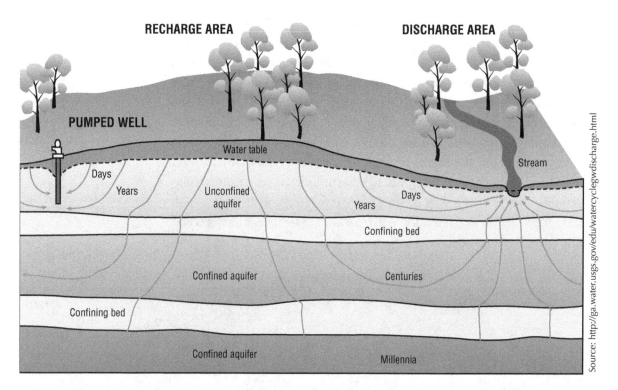

RECHARGE AREA

DISCHARGE AREA

PUMPED WELL

Water table

Stream

Days

Years

Unconfined
aquifer

Days

Years

Confining bed

Confined aquifer

Centuries

Confining bed

Confined aquifer

Millennia

Figure 9.6 Ground Water Flows

Water management is typically focused on two primary issues; *water quantity* and *water quality*. Water quantity is essential to provide sufficient water resources for our various needs, and is determined by the location as well as weather and climate patterns of a region. Water quality is essential to ensure public safety and environmental health, and is determined by the content of the substances dissolved in the water. Contamination of water occurs in a number of different manners, as illustrated in **Figure 9.7**. Contamination (pollution) can be introduced into water sources by *point sources* and *non-point sources*. Point source pollution is contamination discharged into water sources by wastewater treatment plants, factories, or sewers (single *points* where pollution enters the waterway). Non-point pollution includes run-off from agricultural operations, lawns, driveways/parking lots, and from rain entrapping air pollution, and occurs over various distance and time scales. Pollution can also seep into the groundwater through recharge flows.

Water contaminants can be classified into eight primary categories:

1. disease-causing species—including bacteria, viruses, and parasites;
2. inorganic chemicals—including metals, salts, and acids;
3. synthetic compounds—including pesticides, detergents, and solvents;
4. fertilizers—mostly nitrogen- and phosphorous-containing compounds from run off;
5. sediments—including all soil types from land erosion;
6. oxygen-consuming waste—including sewage and animal wastes;
7. radio-active materials—from nuclear fission processes; and,
8. thermal pollution—resulting from heat input from power plants and industry

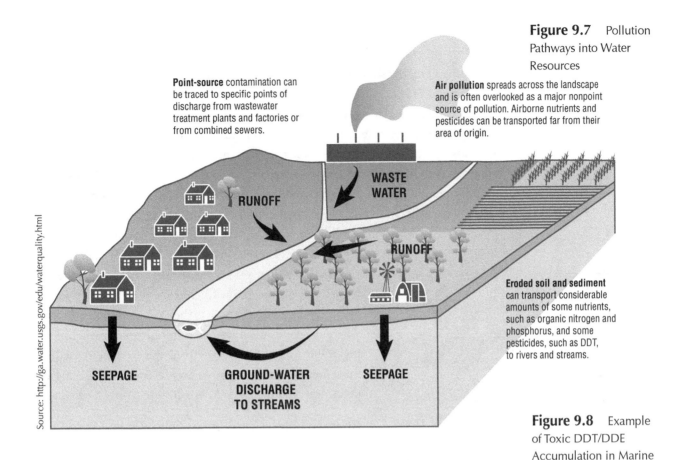

Source: http://ga.water.usgs.gov/edu/waterquality.html

Figure 9.7 Pollution Pathways into Water Resources

Point-source contamination can be traced to specific points of discharge from wastewater treatment plants and factories or from combined sewers.

Air pollution spreads across the landscape and is often overlooked as a major nonpoint source of pollution. Airborne nutrients and pesticides can be transported far from their area of origin.

WASTE WATER

RUNOFF

RUNOFF

Eroded soil and sediment can transport considerable amounts of some nutrients, such as organic nitrogen and phosphorus, and some pesticides, such as DDT, to rivers and streams.

SEEPAGE

GROUND-WATER DISCHARGE TO STREAMS

SEEPAGE

Figure 9.8 Example of Toxic DDT/DDE Accumulation in Marine Ecosystems

Some pollution, like the now-banned toxic pesticide DDT (dichlorodiphenyltrichloroethane) and its relative DDE (dichlorodiphenyldichloroethylene) are persistent in the environment, and concentrate in ecosystems as it moves up the food chain. Mercury from coal burning is similar in nature. For example, as the pollutant enters the water, plankton (at the bottom of the food chain) accumulate the pollutant in their water-bearing tissues. The pollutant becomes increasingly more concentrated as shrimp eat the plankton, fish eat the shrimp, and birds eat the fish. This sequence, presented in **Figure 9.8**, can increase the concentration of the pollutant over a million-fold from the concentration in the water into the concentration in the animals at the top of the food chain. Since humans exist near the top of many food chains, water contamination is of particular importance. In most developed countries, regulated water systems and associated infrastructure are in place to both ensure clean drinking water and minimize down-stream contamination.

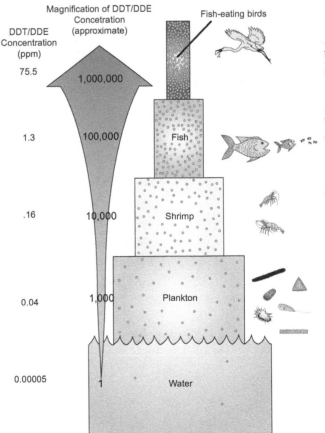

Magnification of DDT/DDE Conceration (approximate)

Fish-eating birds

DDT/DDE Concentration (ppm)	
75.5	1,000,000
1.3	100,000
.16	10,000
0.04	1,000
0.00005	1

Fish

Shrimp

Plankton

Water

Reprinted with permission from the Minerals Education Coalition. From John Christensen-Teri Christensen, *Global Science: Earth/Environmental Systems Science, 7th Edition* (Dubuque, IA: Kendall Hunt Publishing Company, © 2009).

Water Systems

Depending upon the source, water must be cleaned in order to be safe for drinking. In the United States, the EPA regulates public water systems for standards established through the Safe Water Drinking Act (SWDA). Ground water is typically cleaner than surface water, since it has been naturally filtered through the porous rock subsurface. Surface water is exposed to the atmosphere and is more prone to pollution, thereby requiring more treatment. Most water treatment facilities, like that presented in **Figure 9.9**, have similar characteristics. As water from a source is drawn into the system, various coagulation agents are added, which help the suspended solids in the water to settle out in sedimentation tanks. The clarified water from the sedimentation tanks is then processed through filtration systems before it is disinfected with chlorine or other chemicals. The cleaned, filtered water is then pumped into storage, where it resides until we use it in our homes and buildings. The water discharged from the treatment facility must meet specific standards deemed safe for drinking as determined by the SWDA, in addition to any local laws.

After the water is used in our homes or buildings, it is then drained, along with anything mixed with it. Sewers collect this and additional run-off and storm waters to waste water treatment facilities, like that shown in **Figure 9.10**. These facilities first separate the larger debris and grit using screens and settling tanks, before the waste water is introduced into aeration and settling tanks. The aeration tank mixes air with the water to accelerate biological processes, which degrade waste materials into inert compounds. The settling tank allows the heavier components (sludge) to drop out of the water. The sludge contains rich organic matter, some of which is recycled to facilitate the aeration process, and some of which is dried and used as a soil amendment for agriculture operations. The solids-free water is then disinfected with chlorine or other chemicals before it is de-chlorinated and re-introduced into the receiving water body. The water discharged into the receiving water body must meet acceptable standards as established by the SWDA, in addition to any local laws.

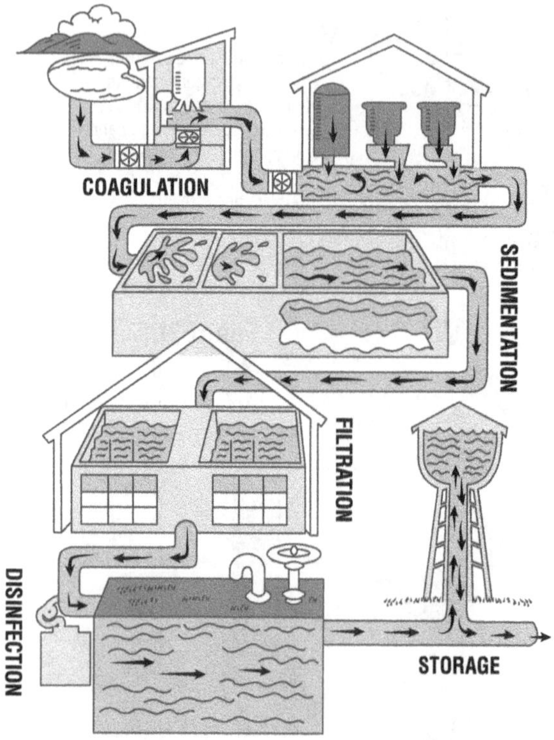

Source: http://water.epa.gov/learn/kids/drinkingwater/watertreatmentplant_index.cfm

Figure 9.9 Basic Components of a Public Drinking Water Treatment System

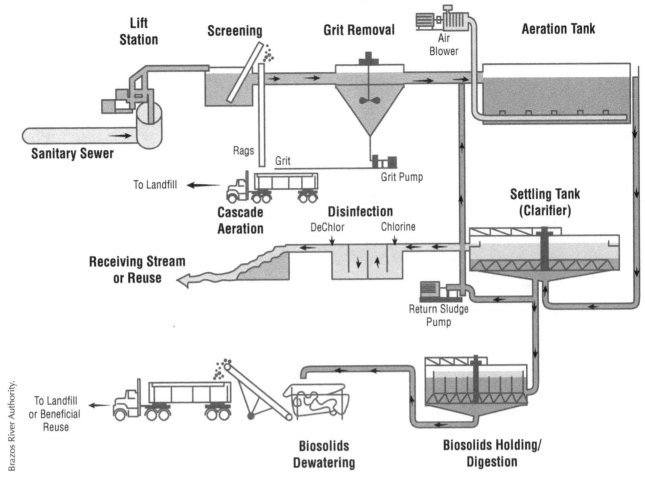

Lift Station

Screening

Grit Removal

Air Blower

Aeration Tank

Sanitary Sewer

Rags

Grit

To Landfill

Cascade Aeration

Grit Pump

Disinfection

DeChlor   Chlorine

Settling Tank (Clarifier)

Receiving Stream or Reuse

Return Sludge Pump

To Landfill or Beneficial Reuse

Biosolids Dewatering

Biosolids Holding/ Digestion

Brazos River Authority.

Figure 9.10 Waste Water Treatment Process

Quiz

(Open Book—Write Answers Below Questions—Show All Work)

1. What are the three (3) primary energy sources within food?

2. What are the six (6) components of the modern food system?

3. What are the three (3) primary soil types?

4. Other than carbon dioxide and water, what are the six nutrients required for plant growth?

5. What factors cause food price instability?

6. How much of the Earth's water is in oceans?

7. How much of the Earth's water is on land surfaces?

8. What are the two primary sources of water contamination?

9. What are the five (5) steps in public drinking water treatment facilities?

10. What are five (5) steps in treating waste water before it is reintroduced into the environment?

chapter 10

Sustainability and Global Society

"People do not change when we tell them they should;
they change when their context tells them they must."

— Thomas Friedman

Definitions, Components, and Metrics

The term *sustainability* is frequently used in today's globalizing society, and therefore, its definition is important. Sustainability describes the ability of a system to maintain a certain state or process indefinitely. In other words, it is the capacity to endure. Since Earth and life on it are in a constant flux of change, the term sustainability must also be put into context. That is, we must discuss sustainability amidst the backdrop of continual change, such as human population growth, development, and its ecological footprint. The term *sustainable development* was popularized in the 1987 United Nations Brundtland Commission on Environment and Development. It stated that,

> *"Sustainable development is development that meets the needs of the present without compromising the ability of future generations to meet their own needs. It contains within it two key concepts: the concept of 'needs', in particular the essential needs of the world's poor, to which overriding priority should be given; and, the idea of limitations imposed by the state of technology and social organization on the environment's ability to meet present and future needs."*

A similar, more-concise description of a sustainable society is provided by Lester Brown, Founder and President of the Earth Policy Institute,

> *"A sustainable society is one which satisfies its needs without diminishing the prospects of future generations."*

These definitions both contain two critical concepts for sustainability; *inter-* and *intra-generational* equity. Inter-generational equity involves equitable access of humans to resources (e.g., food, water, sanitation, energy, and education), regardless of their social or economic status. Intra-generational equity involves the depletion of non-renewable resources and destruction of ecosystem services by the current generation, in effect preventing future generations' access to these resources to meet their needs. The Brundtland definition also explicitly acknowledges challenges inherent in developing countries in terms of relatively unstable social and economic structures and often limited access to technology.

By these measures, our present global human society is not sustainable. First, rampant starvation and poverty reveals lack of inter-generational equity. Second, we are depleting resources and destroying ecosystem services at rates which jeopardize their availability for future generations, and thereby lack intra-generational equity.

In many ways, sustainability is simply an ideal, which may never be fully achieved, yet progress toward it seems a moral imperative. Therefore, we might consider how decisions bring society either toward sustainability or away from it. Perhaps even making sustainability a major factor in decision-making, along with economic benefits and other major factors. To make these assessments, we need metrics to quantify and compare the sustainability of one alternative decision to another. Using careful life-cycle assessment (LCA), we can analyze the use of energy and materials (and their impacts) in every facet of the alternative through its lifetime. Studies have found that for-profit organizations that identify sustainability as a key feature of their operations outperform organizations that do not. This is associated with an organization being more introspective of their decisions, which impact material and energy consumption efficiency, and ultimately increase efficiency and the bottom line: profit.

Sustainability is often measured and discussed using three primary components: environment; society, and economy (**Figure 10.1**). In a basic sense, if the environment is bearable, it will support a society; people have a sufficiently-safe and healthy place to live and meet their needs. If a society is equitable, it will support an economy; people have sufficiently fair-access to make a living and provide for their families. And, if the economy can retain a viable environment, then the system is considered sustainable; the demands on ecosystem services to provide present needs does not preclude the future generations from providing theirs. However, those are rather vague indicators of sustainability (bearable, equitable, and viable).

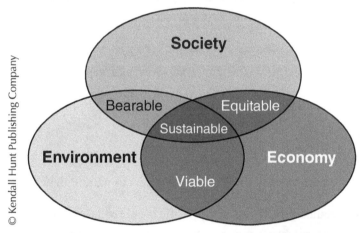

© Kendall Hunt Publishing Company

Figure 10.1 Interaction among sustainability components.

More specific indicators of sustainability for each component are presented in **Table 10.1**. For the environment, we can consider the availability of energy, including food and water resources, and the health of these systems in terms of sustainable supplies. We also can consider the state of the ecosystems that generate renewable resources and absorb our waste streams, and then account for services the ecosystems are able to sustainably afford. One measure of the state of any ecosystem is its biodiversity, or the extent and variety of symbiotic living species. By assessing the state of our ecosystem services and biodiversity, we can determine if the environment can bear the demands of human society. If the environment can support society, we can then consider the state of the society in terms of health, wealth, education, and happiness, and their distribution among the population.

From these measures, we can determine if the society is sufficiently equitable to support an economy. If an economy is sufficiently supported by the environment

Table 10.1 Sustainability components and common indicators

Sustainability Component	Indicators
Environment	energy and water availability, ecosystem services/biodiversity
Society	health, wealth, education, happiness
Economy	gross national product (GPD), debt, employment, stock averages, trade

and society, we can assess its health using numerous measures of debt, employment, stock prices, trade activity, and gross domestic product (GDP—how much a country's economy produces annually). If the economy is healthy, and does not impact the environment and society in unsustainable ways, it is viable. If the environment, society, and economy coexist in harmony and not counterproductively, the entire system is considered sustainable. In a business analog, a sustainable organization is equally concerned with the "3-Ps": profit (economy); public (society); and, place (planet).

One effort in the United States to move the "built-environment" toward sustainability is a program called LEED, for Leadership in Energy and Environmental Design. The LEED program establishes a system for rating the sustainability of new construction, rebuild and remodeling projects in terms of five major categories:

1) site—where the structure is located in terms of ecosystem service impacts;
2) water—consumption, efficiency, and sources of water for indoor and outdoor uses;
3) energy—consumption, efficiency, and sources of energy for HVAC systems and electricity;
4) materials—construction materials and wastes; and,
5) indoor environment—air quality and access to natural light and views.

Through a detailed process, a specific number of credits are assigned to each major category. Other minor categories include additional credits assigned for local and regional sustainability goals. In the end, successful construction/rebuild/remodel projects can be awarded four increasing levels of LEED certification; Certified, Silver, Gold, and Platinum. In addition to being a positive marketing tool, systems like LEED, often result in high-performance, low-cost and long-lasting buildings, thereby promoting sustainable development.

Humans, and all other living species, interact with their environment. When this interaction influences the natural function of the environment or ecosystem in an adverse manner, it is said to have an *environmental impact*. Humans have acknowledged and accepted various environmental impacts throughout history. For example, when buildings are constructed or crops are planted, the land needs to be cleared, excavated, plowed, seeded, etc. Human activities as individuals and societies can therefore cause significant environmental impacts locally, regionally, and ultimately globally, which is especially true with record population and modern technologies. However, these impacts also secure shelter and food, and they have essentially allowed us to become a dominant species on Earth. It is therefore important to understand how environmental impacts can be tolerated by ecosystems in terms of their ability to continue providing essential services.

One generic manner to qualitatively assess environmental impacts is using the "IPAT" formula:

$$I = P \times A \times T$$

I = (Environmental) Impact, P = Population, A = Affluence, T = Technology

This equation multiplies the effects of a technology by the population using the technology, and the affluence of the population (the percentage that have access to the technology). By multiplying (compounding) theses effects, an environmental impact of some technology can be assessed and compared to other technological alternatives. This formula emphasizes the key features contributing to environmental impacts, but only provides a general, qualitative tool to assess environmental impacts.

A more comprehensive assessment of environmental impacts is presented in **Figure 10.2.**[1] Similar to the IPAT formula, impacts of human activities begin with the population, its consumption, and the technologies used in the consumption, or the *Impact Factors*. The human activities responsible for environmental impacts can be categorized into the following *Impact Drivers*:

- agriculture and forests—all the lands for crops, livestock, and timber;
- hunting and fishing—both commercially and for sustenance or recreation;
- urban, industry, and mining operations—providing modern infrastructure, communications, and materials;
- water resources—both ground and surface sources for agriculture, industry, and drinking; and,
- energy and transportation—heating, lighting, electricity, and moving around the planet.

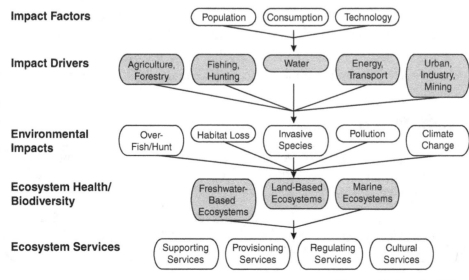

Figure 10.2 Interconnections of human impacts on biodiversity, ecosystems and ecosystem services.

The *Environmental Impacts* themselves can be grouped into:

- habitat loss—debilitation of ecosystem function;
- over-exploitation—over-hunting/fishing or depletion of other renewable resources;
- introduction of invasive species—disrupting naturally-evolved ecosystem symbiosis;
- pollution—emissions of environmental contaminants in amounts beyond the ecosystem's capacity to absorb; and,
- climate change—anthropogenic disruption of Earth's climate system, resulting in rapid warming and associated risks.

We then can assess how these environmental impacts affect the state of *Health/ Biodiversity* within various land-based (terrestrial), freshwater, and marine ecosystems. Finally, we analyze how the impacts affect the *Ecosystem Services* and, ultimately, if and how these services are compromised in terms of supporting the needs of the population. Ecosystem services are often divided into:

- support services—maintaining biogeochemical cycles to support life;
- provisioning services—providing renewed energy and water resources;
- regulating services—like absorbing waste streams and controlling diseases; and,
- cultural service—such as enhancing recreational and aesthetic experiences.

Environmental impacts are often framed in the context of the societal costs and benefits of a particular human activity. This logic is presented in **Figure 10.3**; solid lines indicate direct affects and dashed lines indicate major feedbacks.[2] Beginning with human activity, technology is developed and employed to consume some mass or energy resource and convert it into a more useful one. This process inevitably releases emissions and pollutants into the environment, which causes various environmental impacts. The impacts then are valuated, depending upon the economic and cultural predispositions of the society, and ultimately, the political process within the society establishes public policies (laws, regulations, and standards), which then influences human activity, technology, and thus environmental impacts. Each of the major undergraduate disciplines plays a critical role in this process. The physical and biological sciences provide knowledge on technology function and environmental impacts. Engineering provides the basis for

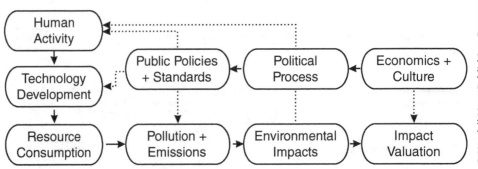

Figure 10.3 Framing Environmental Issues in Society.

technology development, and the arts, humanities and social sciences help develop cultural norms and value systems, which provide context to assess environmental impacts and social value of certain activities and technologies. Agriculture, business, education and political sciences also play pivotal roles in the social, political, and economic systems, which underpin society.

Sustainability can be complicated, but in some ways it can be simple. Consider the sustainable management of waste streams (**Figure 10.4**), for example. Many human activities produce waste, which can be prevented or minimized if alternatives are pursued. When waste must be produced, it can sometimes be reused several times, or recycled into new products. If these more favored options are unavailable, then the residual energy in the waste often can be recovered through incineration (high-temperature combustion/burning) to produce steam for heating or electricity generation. Of course, the least favored option for waste is for permanent storage in landfills or other terminal disposal sites. Adopting this attitude in business and households can significantly improve economics by reducing material and energy inputs and costs associated with waste disposal.

Figure 10.4 Waste management options with most preferred options at top.

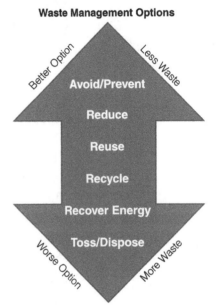

Globalized Societies and Impacts

With improved communications and transport, human society has become increasingly globalized over the past few centuries. **Figure 10.5** presents a map of Earth identifying commercial air traffic (open flights) in a typical day.[3] Comparing this image with Figure 1.1 illustrates the increasingly rapid progression of global human society. While it required tens of thousands of years to migrate to present locations, humans can now travel the globe in a day. While the Agrarian Revolution permitted steadily increasing populations, the Industrial Revolution enabled an exponential increase in global human population. Environmental impacts of human activities have also witnessed exponential increases. In fact, with the advent of rapid fossil fuel burning and discovery of nuclear fission, humans have left a permanent (geological) footprint on Earth, and dramatically affecting its system dynamics. Many scientists have adopted the term "Anthropocene" to describe this relatively new and distinct geological era.[4]

Globalization connects resources with consumption, but often with the exploitation of economically-poor, yet natural resource-rich countries. So while the exponential growth in consumption has supported the rise in economies of the developed world, the developing world continues to catch up. The human development index (HDI—a composite measure of health, wealth, and education) was developed by the UN to compare different regions to one another. The higher the HDI, the healthier, wealthier, and more literate the population (see Figure 1.12).

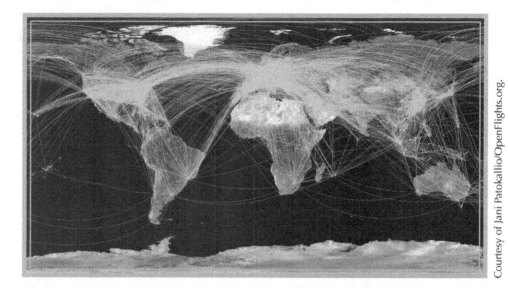

Figure 10.5
Commercial worldwide air traffic[3]

Courtesy of Jani Patokallio/OpenFlights.org.

Figure 1.12 shows the relationship between HDI and ecological footprint, and generally speaking, increasing HDI = increasing footprint. The same is true with HDI and energy consumption, generally more energy = higher HDI. However, in both of these correlations there reaches a plateau of HDI, where additional energy consumption and ecological footprint do not correlate with increased HDI. From this data, it is apparent that people within industrialized regions of the world could live well (maintain a high HDI) while using much less energy, and that people in developing regions need improved access to energy in order to increase their HDI. This concept has been discussed as a combination of *contraction and convergence*, i.e., industrial regions with energy consumption beyond what is required to achieve a high HDI may need to undergo a *contraction* of energy consumption (via efficiency and conservation), while permitting developing regions to pursue a *convergence* of energy consumption with the developing regions.

Science and Informed Citizenry

Science is a method for producing knowledge, which along with philosophy and/or religion can help individuals and society form a sense of purpose and understanding of place, planet, universe, and reality. Science is based on logical principles of careful observation, reasoning, and educated-guessing (proposing an explanation, or generating a *hypothesis*). To test the hypothesis, meticulous experimentation, data collection, and objective analyses of results allow scientists to confirm or revise the hypothesis. After hypotheses have successfully passed through this process, they emerge as scientific *theories* to describe phenomena within the physical or natural world. When theories continue without evidence to contradict their validity, they ultimately evolve into laws, which form the foundation for the scientific understanding of our world. And although there is a formal framework for the scientific method, science is as unique as the scientists who practice it.

Science can be challenging, fun, even sexy, and everyone can be scientific by employing the method in everyday life. For example, if you notice/observe something and wonder how it works, you may form an explanation (hypothesis) to satisfy your curiosity. If you go further to test your explanation (experiment), the results may either support or reject your explanation. You then build experiential, scientific knowledge about the thing you observed. There are numerous examples of using the scientific method in everyday life, from estimating gas costs for a road trip (based on distance and gas mileage estimates) to fixing a seemingly broken flashlight (change batteries) or analyzing diet and exercise (number of Calories eaten vs. number of Calories burned during regular life and activities).

Science also provides a means to be objective in interpreting the vast amount of information we receive on a regular basis. By analyzing the sources of the information and their biases, we may view the information through an objective, scientific lens, to determine if or how to process it into knowledge to inform our own world-view. In this way, people can make more informed decisions which take into account their scientific understanding of the impacts of various actions and their alternatives.

Before realizing the profound change in human activities required to shift the global course toward sustainability, some consensus must be established within local, regional, and ultimately global societies. This can be challenging on all levels. Various governmental and non-governmental organizations (NGOs) have taken on these challenges, to varying degrees of success. Much of the efforts are invested in education and awareness-building within communities and societies to improve understanding of impacts of various activities and how these can be avoided or minimized. A science-literate global citizenry is paramount in these endeavors. To conclude, a number of online resources containing a wealth of information on related topics are listed below; enjoy!

Websites

Climate Change

www.epa.gov/climatechange – US Environmental Protection Agency's comprehensive website on climate change science and impacts

www.globalchange.gov – US global climate change research program website, coordinating among all federal agencies and reporting to the President

www.ipcc.ch – Intergovernmental Panel on Climate Change; official website

Energy

www.energy.gov – US Department of Energy website

www.eia.doe.gov – US Energy Information Agency website

www.iea.org – International Energy Association website

Environment

www.unep.org – The United Nations Environmental Program website

www.epa.gov – US Environmental Protection Agency main website

www.iucn.org – International Union for the Conservation of Nature website

www.nationalgeographic.com – National Geographic Society website

www.noaa.gov – National Oceanic and Atmospheric Administration website

www.learner.org/courses/envsci/ – The Habitable Planet – Annenberg Learner Course

Society

www.un.org – The United Nations main website

www.gapminder.org – Statistics on global society from Swedish NGO

Sustainability

www.worldwatch.org – World Watch Institute website

www.earth-policy.org – Earth Policy Institute website

www.wwf.org – World Wildlife Foundation website

www.epa.gov/sustainability – Environmental Protection Agency sustainability website

References

1. World Wildlife Foundation. 2010. Living Planet Report 2010; Biodiversity, Biocapacity and Development.

2. Rubin, Edward S. 2001. Introduction to Engineering and the Environment: McGraw-Hill.

3. http://openflights.org/demo/openflights-routedb-2048.png

4. Steffen, W., P.J. Crutzen, and J.R. McNeill. 2007. "The Anthropocene: are humans now overwhelming the great forces of nature?" Ambio 36(8): 614–21.

Quiz

. . . .

(Open Book—Write Answers Below Questions—Show All Work)

1. What two (2) key concepts of equity are in most definitions of sustainability?

2. What are the three (3) major components of sustainability?

3. Define and describe the IPAT formula.

4. Describe the four (4) types of ecosystem services.

5. What are the top four (4) preferred methods for waste management?

6. Define and describe the HDI.

7. What are some key indicators for economic sustainability?

8. Describe the three (3) P's of sustainable business.

9. What are some key indicators for environmental sustainability?

10. What are some key indicators for societal sustainability?

Glossary of Terms and Abbreviations

#

3-P's   Profit, People, Planet

A

Activation Energy   Energy required to enable reaction to progress (e.g., spark in combustion)

Aeration   Introducing air through a liquid

AFC   Alkaline Fuel Cell

Agrarian Society   Society based on Agriculture

Agricultural Revolution   Transition to Agrarian societies about ~10-15,000 years ago.

Air Pollution Control   Processing exhaust to remove or reduce air pollutants

Alternating Current (AC)   Electrical charge flow with rapidly changing voltage

Amp   Measure of electrical current / electrical charge flow

Aquifer   Subsurface geological layer containing ground water

Archaea   Distinctive microorganisms that evolved early in Earth's history

Atom   Basic unit of matter—comprised of nucleus containing proton(s) and neutrons orbited by shells of fast-moving electrons

B

Barrel (Oil)   Measurement of oil volume: 42 US gallons, or about 159 liters

Battery   Electrochemical device which converts chemical to electrical energy (some rechargeable)

BCE   Before Common Era—same as "BC"

Bio-capacity   Ecosystem capacity to support population's needs (water, food, waste absorption)

Biodiversity   Measurement of quantity and diversity of living species within ecosystem

Biofuel Liquid biomass, e.g., ethanol, butanol or biodiesel

Biogas Gaseous biomass, e.g., methane or synthetic bio-gas (hydrogen + carbon monoxide)

Biogeochemical Cycle Interconnected cycle within the Earth System involving biosphere, lithosphere, hydrosphere and atmosphere

Biomass Solid organic matter, e.g., trees, dung, algae

Bio-refinery Facility which processes biomass into usable (refined) products.

Bitumen Oil (petroleum) in a semi-solid state, often referred to as "tar sand"

Boilers Devices which boil water into steam using heat (typically from combustion), often referred to as "steam generators"

BTU British Thermal Unit

C

CAFO Concentrated Animal Feed Operation

Capacity Factor Fraction of time that a power plant is operating at full capacity

Carbohydrate Product of photosynthesis—simple organic molecules which form basis for more complex molecules, e.g., sugar

Carbon Element with atomic number 6, basic unit for Organic Chemistry

Carbon Footprint Greenhouse gas emissions emitted via activities of individuals, organizations or countries

Carbon Intensity Greenhouse gas emissions per unit of productivity or energy use, e.g., kg CO_2 emitted per kW-hr electricity consumed

Carbon Monoxide Molecule containing one carbon and one oxygen atom—product of incomplete combustion of carbon and primary air pollutant

Catalyst Material which lowers activation energy for reactions

Catalytic Converter Coverts carbon monoxide to carbon dioxide in exhaust using catalysts

CCS Carbon Capture and Storage

CE Common Era—same as "AD"

CFL Compact Fluorescent Light

Chain Reaction   Series of reactions in which products promote additional reactions

Charge Carriers   Particles carrying electrical charge, e.g., electrons and ions

Chemical Reaction   Process which transforms reactant molecules into different product molecules, either requiring energy (endothermic) or emitting energy (exothermic)

Chemical Weathering   Geological process which changes rock composition, often incorporating atmospheric carbon dioxide into solid carbonates, e.g., limestone

Clean Air Act   US Federal Law which regulates air emissions from stationary and mobile sources

Climate   Long-term measurement of meteorological conditions of an area

Climate Forcing Factors   Large-scale factors, which influence climate, e.g., plate tectonics, Earth's orbit, Sun intensity and GHG emissions

Coagulation   Process in which particles suspended in fluid group-up (agglomerate), which eases their separation

Coal   Fossil fuel comprised of ancient, decomposed organic matter

Combustion   Chemical reaction combining fuel with oxidant to form products, often referred to as burning, or fire

Concentration   Amount of substance within specific volume or mass

Condensation   Phase transformation of vapor to liquid

Conduction (Energy)   Energy transfer through solid

Conductor   Material which exhibits high electrical conductivity

Conservation of Energy   Scientific principle which accounts for energy in all its forms

Conservation of Mass   Scientific principle which accounts for mass in all its forms

Contamination   Introduction of substance into ecosystem beyond its ability to absorb

Contraction (Global Society)   Decreasing carbon intensity by increasing energy efficiency and energy conservation for equity in global resources

Convection (Energy)   Energy transfer by moving molecules, e.g., fan blowing hot air

Convergence (Global Society)   Increasing access to energy and other resources for equity in global resources

Cradle-to-Cradle   Assessment of products from origin and use to recycling into new products

Cradle-to-Grave   Assessment of products from origin and use to disposal

Critical (Nuclear Chain-Reaction)   Sustained Chain Reaction

Critical Mass (Nuclear)   Amount and concentration of nuclear fission material required for sustained chain-reaction

Crude Oil   Raw oil (petroleum) recovered from reservoir

Cyclone (Air Pollution Control)   Device which separates solid particles from exhaust

D

DDT/DDE   Banned toxic pesticides (dichlorodeiphenyltrichloroethane and dichlorodiphenyldichloroethylene)

Demand (Electricity)   Total power requirement of facility from electric utility grid

Deposition   Phase transformation from vapor to solid

Deuterium   Heavy isotope of hydrogen

Dietary Calorie   Measurement of chemical energy within food (kilocalorie)

Direct Current (DC)   Electric charge flow with continuous voltage

Direct Solar Energy   Energy directly converted into thermal or electrical energy

Discharge (Ground Water)   Ground water flow into surface waters

Disinfection   Process to remove or minimize bacterial contaminants

Distillation Tower   Device which separates fluids by differences in boiling temperature—key in refining

Distributed Generation   Electric power generation at many distributed locations versus few centralized locations

E

Earth System   Interconnected system comprised of lithosphere, hydrosphere, atmosphere and biosphere which maintains favorable conditions for life

Earth System—Atmosphere   Thin layer of gas (air) covering Earth's surface

Earth System—Biosphere   Life on Earth in all its forms

Earth System—Hydrosphere   Water on Earth in all its forms

Earth System—Lithosphere   Thin Layer of rock covering Earth's surfaces (ocean floor and continents)

Economy   System of physical and human resources to produce, trade, distribute and consume goods and services

Ecosystem   Community of evolving living and nonliving components linked together by energy flows

Ecosystem Services   Services provided by ecosystem, maintaining biogeochemical cycles, providing resources, absorbing wastes and enhancing experiences

Efficiency   Amount of output versus input

EIA   Energy Information Agency

Electric Current   Flow of electric charge, measured in amps

Electric Power Plants   Facilities which convert mechanical, chemical, nuclear and/or thermal energy into electrical energy

Electricity Grid   Interconnected system connecting electric power plants with electric power consumers comprised of voltage transforming substations, high- and low-voltage power lines and control systems.

Electrolysis   Splitting water (H_2O) into hydrogen (H_2) and oxygen (O_2)

Electromagnetic Spectrum   Series of electromagnetic radiation with varying wavelength, frequency and energy

Electron   Basic unit of the atom—orbits around nucleus in shell-like configuration

Element   Unique atom characterized by number of protons within nucleus

Embedded Energy   Amount of energy required to manufacture and transport products

Energy   Capacity to do work

Energy Conservation   Avoiding energy use

Energy Conversion   Transforming energy from one form to another

Energy Density   Amount of energy within specific volume or mass

Energy Efficiency   Amount of energy successfully converted from less-desirable form to more-desirable form

Energy Forms   Potential (stored) and Kinetic (active) Energy, further categorized into Chemical, Electrical, Gravitational, Mechanical, Nuclear, Radiant (Electromagnetic) and Thermal Energy

Energy Sources   Ultimately from stellar activities, sources can be categorized into: New (Solar Energy); Old (Fossil Fuels); and, Really-Old (Nuclear and Geothermal), and further classified as Renewable or Non-Renewable

Energy Storage   Energy stored in fuels, batteries or Potential Energy (water behind dam), which can be readily converted into usable energy, e.g., electricity

Engines   Devices which employ cycles to convert Thermal or Chemical Energy into Mechanical Energy

Enrichment (Nuclear)   Increasing concentration of nuclear fission materials to reach critical mass

Environment   Features of Earth system in surrounding area (local, regional, global)

Environmental Impact   Disruption of environmental (ecosystem) function due to activity

Environmental Impact Drivers   Population, consumption and technology

Environmental Impact Factors   Agriculture, hunting/fishing, water use, energy/transportation, industry, cities

EOR   Enhanced Oil Recovery

EPA   Environmental Protection Agency

Eubacteria   Distinctive microorganisms that evolved early in Earth's history

Evaporation   Phase transformation from liquid to vapor

Excess Air   Condition in combustion where more air than needed is present (engine running lean)

Excess Fuel   Condition in combustion during which more fuel than needed is present (engine running rich)

Exploration (Oil and Gas)   Process of testing and analyzing geological features in search of oil/gas reservoirs

F

FAO   Food and Agricultural Organization

Fat   Primary component of food with highest energy density

Filtration   Process of separating solid particles from fluids

Flow Cell   Battery with flowing fluids for electrodes

Food   Edible materials containing energy and nutrients to maintain life

Food Groups   Categories of food, typically separated into animal- or plant-based proteins, fats and carbohydrates

Food System   Interconnected physical and human resources to produce, trade, distribute and consume food

Fossil Fuel   Coal, oil or natural gas, decomposition products of ancient organic matter

Frequency   Rate of event, typically measured in hertz (1/second)

Fuel   Material with stored chemical energy to be converted into more-useful energy form

Fuel Cell   Electrochemical device which converts chemical energy into electrical energy

Fungi   Organisms, including mushrooms, which help in decomposition processes within ecosystems

Furnaces   Devices which heat air, typically using combustion or electricity

G

Gas Turbines   Devices which convert Chemical Energy into Mechanical Energy

GDP   Gross Domestic Product

Generators (Electric)   Devices which convert Mechanical Energy into Electrical Energy

GHG   Greenhouse Gas

Greenhouse Effect Effect of retaining infra-red radiation within a system, similar to transparent walls of greenhouses

Grid-Tied Connected with electric utility grid

Ground Water Water residing below Earth's surface

H

HDI Human Development Index

Heat Thermal energy (flow of energy from high temperature to low temperature)

Heat Exchanger Device which transfers Thermal Energy from one fluid into another

Heat Pump Device which actively moves Thermal Energy from one source to another

Horizontal Drilling Technology in which drill moves horizontal to surface after drilled down vertically

Hunter-Gatherer Society Society based on foraging for wild plants and animals

HVAC Heating, Ventilation and Air-Conditioning

Hydraulic Fracturing (Fracking) Technology which introduces high-pressure water and chemicals to fracture oil- and/or gas-bearing rocks to enhance recovery

Hydrocarbons Molecules containing carbon and oxygen

Hydrogen Lightest and most abundant element with atomic number 1

I

IEA International Energy Agency

Incandescent Process of glowing, by heating from electric current

Indirect Solar Solar energy converted into biomass, winds, waves, weather, etc.

Industrial Revolution Transition into industrial (fossil-fuel-driven) society about ~200 years ago

Infiltration (Ground Water) Percolation of surface water into ground water

Insulator   Material with poor electrical conductivity

Inverters   Devices which transform DC into AC electricity or reverse

Ion   Atom or molecule with deficiency or excess of electrical charge

Ionization   Process of adding or removing electrons from atom or molecule

IPCC   Intergovernmental Panel on Climate Change

IR   Infra-Red Radiation

Isotope   Atom with more or less neutrons compared to most stable atom

IUCN   International Union for the Conservation of Nature

J

Joule   Basic SI-derived unit for energy

K

Kilowatt-hour   Common unit of electrical energy

Kinetic Energy   Active energy as in Mechanical, Thermal, Electrical or Radiant Energy

L

Latent Heat   Thermal Energy released or absorbed during phase transformations

LCA   Life-Cycle Analysis

Lead   Element with atomic number 82 and key air pollutant

LED   Light-Emitting Diode

LEED   Leadership in Energy and Environmental Design

Less Developed Countries   Countries with low indicators of socioeconomic status (health, wealth, education, happiness)

Lighting   Radiant Energy within the visible spectrum of frequencies

Line Losses   Electrical energy converted (lost) to Thermal Energy during transport

LNG   Liquefied Natural Gas

Load   Electricity-requiring device

LOSU   Level Of Scientific Understanding

Luminous intensity   Measure of Radiant Energy per frequency per unit area

M

Macronutrients   Materials required in abundance for optimal plant health

Mass   Basic SI unit for measuring matter, kilogram

MCFC   Molten Carbonate Fuel Cell

Methane Hydrate (Clathrate)   Methane molecules trapped within ice lattice cage

Metrics   Units of measurement

Micronutrients   Materials required in trace amounts for optimal plant health

Mole (Amount) 6.022×10^{23}

More Developed Countries   Countries with high indicators of socioeconomic status (health, wealth, education, happiness)

Motors (Electric)   Devices which convert Electrical Energy into Mechanical Energy

N

NASA   National Aeronautics and Space Administration

Natural Gas   Gaseous fossil fuel product of decomposed organic matter

Neutron   Basic component of atom within nucleus

Newton   Basic SI-derived unit of force

NGO   Non-Governmental Organization

Nitrogen Oxides   Oxidized nitrogen molecules product of high-temperature combustion and primary air pollutant

NOAA   National Oceanic and Atmospheric Administration

Non-Point-Source Pollution   Emissions from disperse sources, e.g., storm run-off from parking lots

n-type Semiconductor   Negative-conducting Semiconductor materials (typically via electrons)

Nuclear Fission   Breaking nuclei to create lighter elements and releasing energy

Nuclear Fusion   Combining nuclei to create heavier elements and releasing energy

Nuclear Waste   Bi-products of nuclear fission processes containing persistently radioactive materials

Nucleus   Basic feature of atom, comprised of proton and neutron

O

Off-Grid   Disconnected from electric utility grid

Oil (Petroleum)   Liquid fossil fuel product of decomposed ancient biomass

Order of Magnitude   Scale used to compare large- and small-scale phenomena

OTEC   Ocean Thermal Energy Conversion

Out-Gassing   Emission of volatile gases from solid or fluid

Oxidation Reaction   Reaction which liberates electrons in electrochemical cell at anode

Ozone   Molecule comprised of three oxygen atoms, absorbs ultra-violet radiation in Stratosphere, is an air pollutant from combustion in Troposphere

P

PAFC   Phosphoric Acid Fuel Cell

Particulate Matter (PM)   Solid particles in exhaust which are airborne

PEMFC   Proton Exchange Membrane Fuel Cell

Percolation   Movement and filtration of fluids through porous media

Phase Changes   Conversion of material state from one phase to another

Phase of Matter   State of materials, e.g., solid, liquid, gas or plasma

Photosynthesis   Process of converting Radiant Energy into Chemical Energy

Photovoltaic (PV) Effect   Process where Radiant Energy is converted to Electrical Energy

Piezoelectric   Materials which convert Mechanical Energy into Electrical Energy or reverse

Plutonium   Element with atomic number 94 and radioactive isotopes

p-n Junction   Interface between p- and n-type Semiconductor, critical in PV and LED technologies

Point Source Pollution   Emissions from a single source, e.g., tailpipe or sewer

Pollutants   Contaminants in the environment, materials beyond which eco-systems can absorb

Potential Energy   Stored energy, as in Gravitational, Mechanical, Electrical or Nuclear Energy

Power   Amount of energy per time

Precipitation   Process in which particles settle out from saturated fluid, e.g., rain, snow

Pressure   Measure of force per unit area

Production (Oil and Gas)   Process of recovering oil/gas from reservoirs

Products   Atoms or molecules produced in chemical reactions

Protein   Basic food component and building block of complex life

Protium   Lightest isotope of hydrogen

Proton   Basic component of atomic nucleus

p-type Semiconductor   Positive-conducting Semiconductor materials (typically via electron holes)

Pump Storage   Driving water uphill into reservoir for later use in hydro-electric facility

PV   Photovoltaic (Converting Radiant Energy to Electrical Energy)

PV Arrays   Panels of PV modules

PV Cells   Basic PV unit device

PV Modules   Set of PV cells

R

Radiation   Process of transporting energetic particles or waves

Radiation Forcing   Measure of contribution to energy retention in Earth's atmosphere

Radioactive Decay   Process of emitting radiation during disintegration of isotope

Radio-Carbon Dating   Process of estimating age of organic matter by analyzing concentration of decayed carbon isotopes

Reactants   Atoms or molecules involved in chemical reactions

Recharge (Ground Water)   Transfer of surface water into ground water

Reduction Reaction   Reaction which consumes electrons in an electrochemical cell at cathode

Refinery   Facility which separates crude oil and refines into useful products

Renewability   Measure of resource use versus replacement

Renewable Hydrogen   Hydrogen produced through renewable energy sources

Reservoir (Oil/Gas)   Quantity of recoverable oil/gas in specific geological location

Resistance (Electrical)   Measure of material's inhibition to electrical charge flow

Respiration   Process of ingesting air and exhausting products from biological activities (akin to combustion)

S

Saline   Measure of dissolved salt in water

Salts   Compounds with alkaline and halogen elements, e.g., sodium-chloride (table salt)

Science   Method for producing knowledge comprised of observation, hypothesizing, experimentation, reasoning and repeating

Scrubber (Air Pollution Control)   Device used to remove or reduce pollution from exhaust using liquids

Sedimentation   Tendency of solid particles to settle out of a fluid

Semiconductor   Material with moderate electrical conductivity

Shale   Type of rock comprised of clay and often organic matter

SI   System International (International System of Units)

Smart Grid   Electric utility grid with features promoting increased efficiency, stability and reliability

Society   Group of humans sharing common environment, economy or culture

SOFC   Solid Oxide Fuel Cell

Soil   Complex mixture of air, water and organic and inorganic matter

Soil Types   Categories of soil pertaining to texture and composition, e.g., clay, silt, sand and loam

Solar Electric (Photovoltaic)   Converting Solar Radiant Energy into Electrical Energy

Solar Insolation   Solar Intensity per surface area

Solar Thermal   Solar Energy converted into Thermal Energy

Speed of Light $\sim 3 \times 10^8$ m/s (in vacuum)

Steam Turbines   Devices which convert high-pressure or high-velocity steam into Mechanical Energy

Stoichiometry   Mass balance of chemical reactions (Atomic Accounting)

Stratosphere   Distinct layer of Atmosphere above Troposphere

Stuff   Goods and services enjoyed in modern society

Sub-Critical (Nuclear)   Chain reaction which increasingly diminishes in intensity

Sublimation   Phase transformation of solid to vapor

Sub-Station   Facility where Transformers Change Electric Grid Voltages

Sulfur Dioxide   Oxidized sulfur, product of burning sulfur-containing fuels, which is a key air pollutant

Super-Critical (Nuclear)   Chain reaction which continuously increases in intensity

Surface Water   Waters on Earth's land surfaces

Sustainability   Capacity to endure

Sustainability Components   Environment, society and economy

Sustainability Indicators   Measures within sustainability components to qualify or quantify effectiveness

Sustainable Development   Development which meets present needs without compromising future needs

Symbiotic   Relationship among living organisms with mutual benefits

System   Interconnected components acting together

SDWA   Safe Drinking Water Act

T

Temperature   Physical property of matter which quantifies "hot" and "cold"

Thermodynamics   Study of energy flows and conversions

Thermoelectric   Materials which convert Thermal Energy into Electrical Energy or the reverse

Thermohaline Circulation   Flow of global ocean waters due to differences in temperature and salt concentration

Time   Dimension to order past, present and future and measure duration of events

Transformers   Devices which convert electricity from one voltage to another

Transpiration   Evaporation of water from plants and animals

Tritium   Heaviest isotope of hydrogen

Trophic Levels   Food chain levels within an ecosystem

Troposphere   Lowest level of Earth's Atmosphere

U

UN   United Nations—Group of over 190 countries (Member States)

Undernourishment   Condition of people without foods for sufficient calorie or nutrition requirements

UNEP   United National Environmental Programme

Unit Conversion   Process of converting units of measure

Uranium   Heavy element of atomic number 92, with fissile isotopes

US   United States of America

V

Volt   Basic SI-derived unit of electrical potential

W

Water Quality   Composition of water resource, measured in concentration of non-water components.

Water Quantity   Amount of water resource, measured in volume

Water System   Interconnected physical and human resources for extraction/collection, trade, distribution, use and treatment of water

Water Table   Level of ground water (below surface) in an area

Water Treatment   Process to remove or reduce contaminants within water

Watt   Basic SI-derived unit of power (energy/time = joule/second)

Wavelength   Measurement of distance between waves (length)

Weather   Short-term meteorological conditions in an area

Work   Measurement of Mechanical or Electrical Energy, force multiplied by distance or voltage multiplied by current and time

WRI   World Resources Institute

WWF   World Wildlife Foundation